William Fuller

Architecture of the Brain

William Fuller

Architecture of the Brain

ISBN/EAN: 9783744679220

Printed in Europe, USA, Canada, Australia, Japan

Cover: Foto ©berggeist007 / pixelio.de

More available books at **www.hansebooks.com**

BY

WM. FULLER, M. D.

Formerly Demonstrator of Anatomy, McGill College; Professor of
Anatomy, Bishop's College, Montreal, Canada; Member
of the American Medical Association, etc.

ILLUSTRATED.

GRAND RAPIDS, MICH.
1896.

Press of
Seymour & Muir Printing Co.,
Grand Rapids, Mich.

Engravings by
Grand Rapids Engraving Co.

INTRODUCTION.

In the following description of the Central Nervous System it is not intended to rehearse minute details which can be found in the many excellent works upon anatomy, but to place in view in as clear and concise a manner as possible the general architecture of the central nervous system, and to trace the relation and continuity of its parts. When a general knowledge of the structure of the brain is acquired by the student a useful and practical step is gained, because he will not only be able to describe the situation of a lesion and understand the descriptions made by others, but he will be in a situation to intelligently discuss the functions of its parts, and is prepared to work in the field of discovery. The incidental assignment of function to any part of the nervous system in these pages is intended to assist the memory and awaken enquiry. Nothing stimulates the observation to the same extent as the entertainment of a theory, to be demonstrated or corrected by careful observation of facts which fall under one's own notice, or by those which can be obtained from other reliable sources.

The following description has been made from dissections by the author, and has been carefully verified by a comparison of longitudinal and lateral dissections, and by sections, all of which agree in proving the correctness of the representations herein described.

These dissections have been permanently recorded by castings in plaster, and the sections have been photographed and plates made, so that both the casts and the plates can be repeated at pleasure, thus enabling any who may take an interest in the anatomy of the brain to acquire a knowledge of it with ease.

A cast is much superior to a picture, since it represents correctly the size, relation and direction of the parts in a much plainer manner than can be done by a plate.

The plates introduced for illustration are drawn from the casts of dissections and refer to them.

In order to bring the means of acquiring a knowledge of this important department of medical science within the reach of all, an Anatomical Company has been formed to reproduce the casts and plates at a very moderate price.

THE MEMBRANES OF THE BRAIN.

The Central Nervous System, or the Cerebro-Spinal axis, is contained in the cavity of the cranium and spinal canal. This cavity is lined by a tough fibrous membrane, the dura mater, which is closely adherent to the inner surfaces of the bones of the skull, especially at the sutures, base, and the margin of the foramen magnum, but is loosely attached by fibrous tissue to the bones forming the spinal canal. Its internal surface is smooth, covered by a serous surface which is the outer wall of the arachnoid cavity, and it is perforated by numerous openings corresponding to the foramina of its bony enclosure. These openings are for the transmission of nerves and blood vessels, and the membrane is prolonged **outwards, to be** continuous with their sheaths and with the periosteum of the skull. As an instance, the dura mater forms a sheath for the optic nerve, expands to form sclerotic coat of the eyeball and lines the walls of the orbit. The dura mater contains between its layers the meningeal arteries and channels for the conveyance of venous blood, called sinuses. The falx cerebri and the tentorum cerebelli are projections of the dura mater extending into the fissures of the brain, respectively between the hemispheres of the cerebrum and between the occipital lobes and the cerebellum. The use of these processes are to sustain the parts of the brain in their proper position.

The arachnoid is a thin fibrous membrane which surrounds the brain, spinal cord and **the roots of the** nerves. It covers the convolutions of the brain, but does not dip into the sulci between them. It is somewhat loosely attached to the surface of the brain, but closely adherent to the pons, medulla and spinal cord, as well as to the roots of the nerves which it encloses and supports between their origin and the foramina of the dura mater before mentioned. The intervals between the arachnoid membrane and the pia mater beneath are called the sub-arachnoid spaces. These communicate with each other and with the ventricles of the brain and in certain pathological conditions are the seat of effusion. The arachnoid is continuous with the inner serous surface of the dura mater enclosing the arachnoid cavity. Upon each side of the spinal cord, between the roots of the nerves and throughout its whole length, this membrane sends out delicate projections which are attached to the inner surface of the dura mater, the function of which is to preserve the position of the spinal cord in the spinal canal. Together they are called the ligamentum dentatum.

The pia mater immediately invests the nervous substance of the brain and spinal cord, is the vascular membrane, and is made up of an intricate network of blood vessels of various sizes, which break up into numerous minute vessels that penetrate the nervous mass and are distributed to it. In a section of the brain these vessels are divided and are seen as points or specks on the surface of the section, and are called puncta vasculosa. Those blood vessels which are distributed to the grey masses, or nerve centres within the brain, are large and numerous, while those in the white matter are smaller and fewer in number. Projections of the pia mater extend through fissures into the ventricles, or cavities within the brain, with which they will be described. Attached to these projections of the pia in the ventricles are several close capillary networks, one for each ventricle, which are called the choroid plexuses and which will also be described with the ventricles of the brain. The network of blood vessels constituting the pia mater is held together and supported by fine filaments of fibrous tissue attached on the outer side to the inner surface of the arachnoid membrane, and on the inner extending into the substance of the brain with the framework of which it is continuous. The arachnoid membrane before described is a condensation of the fibrous tissue of the pia mater over the surface of the brain covered externally by a serous surface, in the same manner as the internal surface of the dura mater is covered internally by a membrane of the same serous character. These serous surfaces are continuous, therefore it may be said, that in reality there are but two membranes of the brain, viz., the dura mater and the pia mater, each having a serous surface which together enclose a serous cavity.

To sum up: The brain is supplied with a dura mater for protection, the arachnoid to accommodate its movements and the pia mater to provide for its nourishment.

THE CEREBRO SPINAL AXIS.

The mass of nervous matter included under this heading is commonly referred to as the brain and spinal cord. The brain is that portion of it which is contained within the cranial cavity, and the spinal cord is the part suspended within the spinal canal. In this description we will view the axis as an isolated object, and the divisions of it made for the convenience of description, will be marked by distinctions upon the axis itself, without any reference to the relations of its parts to those of any other part of the body.

After giving a general description of the structure of any part, we will direct the attention to the natural order of dissection from above downwards, as represented by the castings before referred to, which are intended to accompany and illustrate these pages.

When the anatomy of the spinal cord has been reached in the order of description and studied, its various columns will be traced upwards to their associations above, by which a complete review of the whole will be brought before the mind and its parts associated into the formation of a structure.

Divisions of the Axis.

A very proper and convenient division of the cerebro spinal axis is the *cerebrum*, *cerebellum* and *spinal cord* with their respective conections, and in this description we will follow the divisions commonly in use though without strict adherence to old boundaries. We will, for instance, say that the cerebrum is all that portion of the brain above the pons Varolii, including the hemispheres their peduncles and commissures.

Cerebrum.

The cerebellum and pons will be described together, on ac- **Cerebellum.**
count of the close association of their parts and the necessity
of maintaining the analogy between this structure and that
of the cerebrum. The medulla oblongata will be outlined, **Spinal cord.**
as that portion of the spinal cord which extends from the
lower border of the pons Varolii to the lower part of the
decussation of the anterior pyramids of the medulla, below
which point is the spinal cord proper.

Viewed as a whole the cerebro spinal axis presents two
lateral halves, united by commissures and divided in the middle before and behind by a continuous but modified longitudinal fissure, which in front is called the anterior and
behind the posterior median fissure.

THE MEDIAN FISSURE.—The median fissure divides the
axis into lateral halves and has different names applied to it
in the different localities in which it lies. In front of the
spinal cord it is called the anterior median fissure, which is Ant. median fissure
partially interrupted at about an inch below the pons Varolii Figs. 7 9.
by the decussation of the anterior pyramids of the medulla oblongata. Unless the groove upon the front of the pons Varolii
which lodges the basilar artery may be called a continuation of Basilar groove.
the anterior median fissure, this fissure is completely inter- Figs. 9 27.
rupted to the extent of about one and half an inch by the
transverse fibres of the pons. Above the pons, this fissure
is broad and expanded between the peduncles of the cere-
brum, where it is called the interpeduncular space, in front of Interpeduncular
which, at its termination in the great longitudinal fissure of space. Compare
the cerebrum, it is interrupted by the optic commissure. Figs. 2 7 9.

The sides of the interpeduncular space are connected in
front with the commencement of the fissures of Sylvius, and

NOTE.—The vertical sections are numbered separately to indicate the situation of each cut
up on plates 25 and 26 and are marked by a * thus (25*). The general plates are referred to by simple
numbers.

behind with the lateral extremities of the great transverse fissure.

Great longitudinal fissure. Figs. 1-2-17 24 26.

The great longitudinal fissure separates the hemispheres of the cerebrum from each other, it averages about two and one-half inches in depth, and at its bottom is the corpus callosum, the great commissure of the cerebrum which unites the hemispheres. Behind the posterior extremity of the corpus callosum this fissure joins the great transverse fissure, which it intersects in the middle of the latter. Beneath the posterior extremity of the corpus callosum both communicate with the cavity of the third ventricle. A process of pia mater, called the velum interpositum, passes through this opening or fissure, the fissure of Bichat, into the third ventricle. A longitudinal groove separating the tubercles of the corpora quadrigemina, and lower down a depression upon the upper surface of the valve of Vieussens, mark the continuation of the median fissure, which is now interrupted by the cerebellum. The fissure is indicated upon the cerebellum by a notch in front and one behind, the incisura anterior and posterior of the cerebellum. Below the cerebellum, the posterior median fissure begins at the lower extremity of the fourth ventricle, and extends throughout the length of the spinal cord. The posterior median fissure of the spinal cord is deeper and narrower than the anterior, and its sides are more difficult to separate.

Intersection with transverse fissure Fig. 26.

Velum interpositum. Fig. 5.

Groove. Fig. 8.

Incisura. Figs. 27-29.

Posterior median fissure. Figs. 8-10.

THE VENTRICULAR SYSTEM.—In the centre of the mass throughout its whole length is a cavity or ventricle which is also modified at different points along its course. In the spinal cord it is a slender canal which, behind the medulla oblongata, connects with the posterior median fissure by an interruption of the posterior commissure of the spinal cord. At this point, on account of the lateral separation of the posterior columns of the cord, the minute canal expands into a broad cavity, the

Central canal of spinal cord. Figs. 8-10.

fourth ventricle, which is situated **behind** the upper half **of** the medulla oblongata and the pons Varolii, and is beneath the cerebellum. Above the pons, the cavity of the fourth ventricle is contracted into a canal about the size of a small quill, the **iter**, which is about five-eighths of an inch in length, **lies** beneath the optic lobes and between the crura of the cerebrum, and connects the third and fourth **ventricles of the** brain.

The iter terminates **above** in a vertical slit, the third ventricle, situated between **the opposed sides of the thalami optici, or basal ganglia of the cerebrum. The third ventricle reaches below to the base of the** brain, and above extends laterally outward **over the thalami to the edges of the fornix, beneath the margins of which it is continuous with the** lateral **ventricles** of the hemispheres. The lateral ventricles are large cavities, one in the center of each hemisphere, prolongations of which extend into the lobes of the hemispheres, and are called the anterior, middle and posterior cornua of the lateral ventricles.

Thus it may be said that the cerebro spinal axis is a hollow tube of nervous matter divided into lateral halves which are separated by a fissure and united by a commissure. The cavities will be again described in connection with the parts that enclose them.

Fourth ventricle compose figures 8-10-13-15-20.

Iter. Figs. 10-15 26-27.

Third ventricle. Figs. 6-8-10-15-26

Lateral ventricles. Fig. 4.

THE CEREBRUM.

THE SURFACE.

Separated from the rest of the mass of the Central Nervous System and viewed, the cerebrum is much the largest part, being about eight times the weight of the cerebellum, and thirty times that of the spinal cord. It averages seven inches in a longitudinal and five inches in a transverse diameter. It is convex upon its upper surface, and its sides seen from above, present an oval contour.

Superior surface. Figs. 1-24 26.

Turned over, its under surface in a general aspect is oval and flat. This surface presents three lobes and two surfaces, and is divided into lateral halves by a continuation of fissures and depressions.

Inferior surface Figs. 2-7.

The anterior or frontal portion is a flattened surface divided in the middle by the anterior median fissure. The marginal convolutions, which form the sides of the fissure, are elevated above the rest of the surface, and the parts upon each side of the elevation are triangular and slightly concave, corresponding to the upper surfaces of the orbital plates of the frontal bone, upon which they rest in their normal position in the cranial cavity. On the sides of the median fissure, and removed half an inch from it, are two other fissures, one on either side, which run parallel to it and lodge the olfactory bulbs and nerves.

Frontal surface.

Olfactory nerves, bulbs and fissure. Figs. 2-7.

The temporal lobes project above this surface behind like large tubercles, one on each side, and are separated from each other by a considerable interval. They are divided from

Temporal lobes.

the frontal surface, or lobes, **by the fissures of Sylvius which arch around and constrict their bases.**

The posterior, or occipital lobes, are directed backwards, are pointed posteriorly, and are separated from each other by the posterior median fissure, which in this situation, longitudinally, is about three inches in depth, running forward as far as the posterior **border of the corpus callosum.**

The contour presented by a view **of the under surfaces of the temporal and occipital lobes** is **oval, concave from before backwards and from side to side, forming a shallow cup, in** the center **of which is an oval opening. This opening is the** hilus **of the cerebrum** which **transmits the cerebral peduncles in front, and behind lodges the** optic lobes, or corpora quadrigemina. In front **of this** opening, between the temporal **lobes, is an** interspace, the **floor of which** contains the **optic** commissure, infundibulum, **corpora** albicantia **and posterior perforated space.** These parts rest upon the body **of the sphenoid bone. The** posterior median fissure **extending** backward **completes the division** of the surface into **lateral halves.**

The anterior portions of this surface, or the temporal lobes, **are lodged in the temporal** fossæ **of the skull, and the occipital lobes rest upon the tentorium cerebelli.** This surface behind forms the upper boundary of the extra ventricular portion of the great transverse fissure or fissure of Bichat.

The cerebrum is divided into lateral halves, called **hemispheres, by the great longitudinal fissure, which** includes the **anterior and posterior median fissures that** separate the frontal **and the occipital lobes. The** hemispheres are united at **the base of the cerebrum, and by** several commissures the largest of which is the corpus collosum **or the great transverse commissure of the** cerebrum.

Fissure of Sylvius.
Figs. 2 7-16-21.

Occipital lobes **and** posterior **median** fissure.

Temporo-occipital surface.

Hilus of cerebrum and of the hemispherical ganglion. Figs. 2 17

Fissure of Bichat in Fig. 2 bounds the hilus on its sides and behind. The ventricular extremity, Figs 4 and 5. Its inferior extremity is seen in Fig. 17. It is often called the transverse fissure.

Corpus callosum.
Figs. 3 8 23 26.

In general contour and markings the lateral halves of the cerebrum are alike, but upon closer inspection, the convolutions and fissures of one hemisphere do not correspond with those of the other, and in a comparison of any number of brains, there cannot be found any exact resemblance in the hemispheres of the same side.

Compare convolutions and fissures Figs. 16 with 24 and 17 with 26.

There are, however, certain distinct fissures and convolutions which enable the anatomist to map out the surface of the brain, and to define localities which may be compared with those of other brains. The whole surface of the brain is divided into general regions which correspond to and are named after the bones of the skull which cover them. Thus we speak of the frontal, parietal, occipital, temporal, temporo-sphenoidal, orbital and insular regions of the brain, and that which lies in apposition with its fellow within the longitudinal fissure, as the inner or fissural region.

See description of hemispherical ganglion.

The convolutions upon the surface of the cerebrum will be described later on with the hemispherical ganglion.

STRUCTURE OF THE CEREBRUM.

In a general description of the anatomy of the cerebrum, it may be said to arise by two peduncles, the parts of which are derived in front from the pons Variolii, and behind from the cerebellum. Each peduncle is about five-eighths of an inch in length, about one inch in breadth from before backwards, and about three-fourths of an inch in thickness in a lateral direction. The peduncles are separated from each other in the median lines by a thin layer of antero-posterior fibres called the raphe.

Peduncles of cerebrum. Figs. 2 7-8 9-10-26 27.

Raphe of peduncles. Figs. 26 27-31.

One-third of the distance from the posterior median line, on the outer surface of each crus, is a vertical depression leading upwards from the pons to a tubercle, the internal geniculate body. In front of this depression is a broad band of coarse

Geniculate bodies. Figs. 7-8 9 12 13 and vertical sections Nos. 20-21- (G. G.)

THE CEREBRUM.

fibres, the crusta, which is directed upwards and backwards to the lower margin of the optic tract that marks the upper boundary of the cerebral peduncle. Behind the vertical depression is an oblique tract of finer fibres having the same direction as those of the crusta and extending upwards to the testes, and to a horizontal tract which is directed outwards from the testes to the internal geniculate body, the brachium posterius. The oblique tract behind the depression arises apparently from the pons Varolii, but subsequently will be traced to the anterior column of the spinal cord. This tract is depressed below the level of the crusta and overlaps a large band of horizontal anterioposterior fibres derived from the cerebellum, the processes e cerebello ad testes (a misnomer).

The *processus* (a name which we will continue to use in place of the former) of one side is connected with that of the other along their internal margins by the valve of Vieussens which, together with the processes, forms the roof of the fourth ventricle. The crustæ diverge as they pass upwards from the pons and are separated in front by a wide interval, the interpeduncular space. This interval between the crustæ in front is triangular in shape, pointed below, and is about a half an inch in depth above the pons. Two round white eminences, the corpora albicantia, are situated between the peduncles about half an inch in front of the pons Varolii. The third nerves spring from the inner margin of the crustæ immediately above the upper border of the pons. The crustæ are about a quarter of an inch in thickness, and the nervous mass enclosed and supported by them, composed of numerous tracts and masses of white and gray matter, is called the tegmentum. The crustæ are the motor tracts and expand above bodily into the internal capsules of the cerebral hemis-

Side notes:
Crusta. Figs. 2-7 8 9 10 11 12 15 21-22 23 27.

Vertical depression on the side of the peduncle. Fig. 9.

Nates, testes, brachis, pineal gland. Figs. 6 8 9 10 13 14 26 27.

This oblique tract is the fundamental root zone. Figs. 8 9 11 12 13.

Processes. Figs. 8 9 10 11 12 13 14-15 27 28 21* to 23*.

Corpora albicantia, third nerves, tuber cinereum and infundibulum. Fig. 9.

Crustæ and tegmentum. Fig. 32.

Expansion of crustæ. Figs. 13 21 19.

pheres, passing directly through them to reach the motor areas above upon the sides of the surface of the brain. The tegmentum contains the sensory tract, and, with one exception, the fillet, its fibrous structures are interrupted by masses of grey matter before entering the internal capsule to be distributed to the surface of the brain. In a section of the peduncles the tegmentum consists of all that portion of the surface except the divided tracts of the crustæ, which are situated on the sides and the front of the peduncles.

Tracts of tegmentum. Figs. 10-15.

Tegmentum. Fig. 22.

The anatomy of the peduncle will be examined more closely later, but at present it is necessary to mark the distinction between the crusta and the tegmentum, because these divisions of the crus cerebri or peduncle are frequently referred to in works upon anatomy.

The internal capsule of each hemisphere is derived from the expansion of a crus or peduncle of the cerebrum. The fibres of the crus cerebri pass upwards and outward to about the middle of the hemisphere, where they decussate with the fibres of the corpus callosum, external to and above the outer border of the lateral ventricle. Beyond the decussation of these tracts, which extends from before backwards throughout the length of the outer margin of lateral ventricle, the fibres of the corpus callosum and internal capsule intermingle with those of the longitudinal commissures of the hemisphere and constitute its white substance, or the corona radiata.

Internal capsule. Figs. 4 to 14, 18 to 22, 10* to 19*.

Decussation of corpus callosum and internal capsule. Figs. 3 8-18*- 19*.

Corona radiata and puncta vesculosa. Figs. 3 to 6.

The corpus callosum is the great transverse commissure of the brain. It forms the roof the lateral ventricles, and uniting upon its sides with the fibres of the internal capsules, a triangular space is enclosed containing the ventricles and the internal basal ganglia.

Triangular space enclosed by corpus callosum and internal capsule. Figs. 8, 16* to 19*.

A vertical section across the middle of the brain shows the internal capsule below its junction with the corpus cal-

THE CEREBRUM.

losum to be in relation with internal and external masses of grey matter, the basal ganglia. Those on the inside, are the caudate nucleus, and below it the thalamus opticus; upon the outside of the capsule are the three divisions of the lenticular body, the corpus lenticularis which is bounded externally by the **external capsule**, claustrum and insula successively. Above the decussation of the internal capsule and the corpus callosum is the white substance of the brain, the **corona radiata**, composed of three layers; an external or sub-convolutional, a middle or the layer of the longitudinal commissures, and an internal or capsular layer which immediately surrounds the ventricle.

The mass of white matter constituting the corona radiata is surrounded by a sheet of grey matter which extends along the surface of the convolutions and into the depths of the sulci upon the surface of the cerebrum. This lamina of grey matter covering the hemispheres constitutes the hemispherical **ganglia** and is the seat of consciousness, memory and mind.

The foregoing **general** description has prepared us to consider each part of the cerebrum more particularly as it appears in the order of dissection.

[margin notes: Lenticular nucleus. Figs. 9, 12, 17*. Layers of corona radiata. Figs. 20-21-25*. Hemispherical ganglion. Figs. 16 17.]

DISSECTION.

Corpus callosum.
Fig. 3 B

By separating the sides of the longitudinal fissure between the hemispheres of the cerebrum, there is brought into view a broad band of transverse fibres, the corpus callosum, which connects the hemispheres. Enclosing the corpus callosum behind, above and in front is a long continuous convolution, the gyrus fornicatus, which overlaps it on each side as far as the middle line. The fissure between the corpus callosum and the gyrus fornicatus is about five-eights of an inch in depth, extending in a lateral direction from the bottom of the longitudinal fissure, and is sometimes called the ventricle of the corpus callosum.

Gyrus F. Figs. 17-26.

Crest. Figs. 3-8.

If the convolutions above the corpus callosum are gently lifted and torn outward, they will break externally from an elevated crest, one on each side, which is the line of decussation of the corpus callosum with the internal capsule as it ascends from the crus cerebri. These crests extend longitudinally along the middle of the hemispheres to near their extremities, are more elevated in the center than at each end, and are more widely separated behind than in front. The crests project upwards as longitudinal ridges that form the outer boundaries of the corpus callosum. The upper surface of the corpus callosum is concave from side to side forming a broad longitudinal groove, or trough about four inches in length and two inches wide, situated in the middle of the cerebrum.

External to the corpus callosum on each side is the white substance of the brain, the corona radiata, divided in front

DISSECTION. 19

and behind into two parts by the anterior and the posterior median fissures. It is fringed by a thin layer of grey matter upon its outer border, and the broken surface shows the ends of torn blood vessels, the puncta vasculosa.

Puncta vasculosa. Fig. 2

The whole surface exposed by breaking the upper part of the cerebrum to a level with the corpus callosum is oval in shape, and convoluted on its margin by fissures. One on each side is deep, and situated behind the middle of the hemisphere, the posterior extremity of the fissure of Sylvius. Behind the corpus callosum is the exposed upper surface of the cuneus lobe of the brain, the center of vision, and the convolution immediately behind the fissure of Sylvius is the center of word memories.

Cuneus. Figs 3 17 20.

To illustrate the structure of the corona radiata a triangle has been figured upon the right side of figure 3, which serves to elucidate the phenomena of word blindness, word deafness, amnesia and aphasia, also another triangle in front, to illustrate cerebral automatic action. The external side of the triangle represents the outermost layer of the corona radiata which is composed of fibres passing between adjacent lobes and convolutions. These are called the internuntial fibres, and in the illustration the tract is represented through which the name is suggested by the object, and vice versa, and the fingers are set in motion automatically by aural and visial impressions.

Internuntial fibres. Figs. 3 and 18 19 22.

The internal and anterior sides of the triangle represent the fibres of the external longitudinal commissure which associate the memory centres behind with the seat of the intellect in the anterior lobes, and with the motor centres of the central region of the brain. The former for the production of conscious and rational, and the latter for subconscious or automatic action. It is easy to determine the results produced upon memory, mind and action by the destruction of a nerve centre, or the interruption of any of these tracts by accident or disease.

The internal layer of the corona radiata is that next to and including the crest or decussation of the corpus callosum and internal capsule. This layer of the corona lies close to the walls of the lateral ventricle previous to its distribution. External to this layer, the fibres of the several layers are mingled into a network, the formatio reticularis of the cerebrum. The internal layer is composed wholly of fibres from the corpus callosum and the internal capsule, and as they pass toward the surface of the brain are disposed in horizontal laminæ between which the fibres of the longitudinal commissures are inserted in their passage from before backwards. The outermost layer of the corona radiata is a close network of fibres passing in all directions and therefore in dissection breaks into short pieces immediately beneath the convolutions. The three layers of the corona may with propriety be called, from within outwards, the capsular, commissural and internuntial layers. These layers are exhibited in plate No. 25 of the vertical sections.

Longitudinal commissures, superior or commissure of gyrus, inferior or fornix, and external. Figs. 3 4 9 18 to 21.

THE LONGITUDINAL COMMISSURES.—The longitudinal fibres, or commissures of the hemispheres, are probably scattered through the entire mass of the corona, but in some situations they can be demonstrated with ease and the fibres separated for long distances. A tract running parallel with the bottom of the fissure of Sylvius, beneath the insula, is of considerable size, and is called the external longitudinal commissure. This commissure is represented in the triangle just described, and terminates in the internal capsule in front of the lenticular body. Another commisure is situated in the gyrus fornicatus upon the inner side of the crest and

External longitudinal commissure. Figs. 19 20 21.

resting within the outer margin of the corpus callosum. It is the superior longitudinal commissure, or the commissure of the gyrus fornicatus.

Superior Longitudinal commissure. Figs. 3 23.

THE RAPHE AND STRIÆ LONGITUDINALES.—In the middle line upon the upper surface of the corpus callosum is a slight linear longitudinal depression, the raphe, and upon each side, and parallel to it, are two longitudinal white lines, the striæ longitudinales, or the nerves of Lancisi. The anterior fibres of the corpus callosum are arched forward into the frontal lobe, the middle fibres are transverse, and the posterior form a broad arch behind and are directed backwards beneath the posterior cornua of the lateral ventricles to be distributed to the occipital, basalar and temporal convolutions. The fibres of the corpus callosum decussate continuously with those of the internal capsule above the outer margin of the lateral ventricle and the roof of the posterior cornu, forming the crest. The posterior margin of the corpus callosum is thick, is inverted beneath the posterior edge of the fornix, to which it is attached and is called the splenium.

Raphe and striæ longitudinales. Fig. 3.

Decussation over the roof of posterior cornu. Fig. 27.

Splenium, Figs. 4- 5 26.

The splenium forms a part of the upper boundary of the intraventricular portion of the transverse fissure of Bechat.

The anterior margin of the corpus callosum is arched downwards to the base of the brain, where it divides into two peduncles that diverge outward obliquely across the anterior perforated spaces of opposite hemispheres to terminate in the anterior part of the temporal lobe upon each side. At their angle of divergence they bound the lamina cinerea in front. The downward deflection of the corpus callosum is called its genu or knee.

Anterior peduncles of corpus callosum, Fig. 7.

Lamina cinerea. Figs. 9 11 12.

Genu corpus callosum. Figs. 4 5-6- 26 9*.

A vertical section of the corpus callosum shows that it is about double in thickness in front and behind than in

Fig. 20.

the middle, and in separating its fibres, the inferior layer is observed to be turned downward at its lateral extremities into the caudate nucleii, while the upper fibres pass upward into the crest on either side. The under surface of the corpus callosum forms the roof of the lateral ventricles and is attached below along the median line from before backwards, behind to the fornix, and in front to the septum lucidum. Its attachment to these parts bounds the lateral ventricles internally and separates them from each other.

Septum lucidum.
Figs. 4 5 6 8 10-20 (S. L.)

The removal of the corpus callosum exposes the lateral ventricles, except the middle and posterior cornua which are uncovered by a section of the temporal and occipital lobes. The cavities thus exposed together present a broad depressed surface between the internal capsules. These project upward and outward, and form its outer boundries on each side for two-thirds of the distance in front.

Lateral ventricles.
Fig. 4.

Projections of the lateral ventricles extend into the anterior, middle and posterior lobes of the cerebrum, and are called the anterior, middle and posterior cornua of the lateral ventricles. The anterior cornua are directed forward and outward, the middle backwards, outwards, downward, forward and inward, encircling the crura of the cerebrum and upon the floor of each is a rounded longitudinal eminence, the hippocampus major which terminates below in a broad convoluted extremity, the pes hippocampi. The posterior cornu is directed outward, downward, backward and inward, and projecting from its internal wall is a longitudinal eminence, the hippocampus minor. An oval eminence is situated externally at the junction of the middle and posterior cornua and is called the pes accessorius, or eminentia collateralis. Behind the reflection of the internal capsule, as it bends downward into the roof of the middle cornu, is a broad groove, directed outwards and downwards,

Cornua of the lateral ventricles.
Fig. 5.

Hippocampus major, minor and pes accessorius.
Fig. 5.

Broad groove.
Fig. 4 5.

which is the common communication between the middle and posterior cornua with the body of the lateral ventricle.

The parts seen upon the floor of a lateral ventricle, from before backwards, are the caudate nucleus, the tænia semicircularis, a portion of the thalamus opticus, the choroid plexus, corpus fimbriatum and the fornix. Between the anterior pillars of the fornix and the anterior extremity of the thalamus is an oval opening through which the lateral ventricle communicates with the third ventricle, and with the lateral ventricle of the opposite side. This opening transmits the choroid plexus of one ventricle to be continuous with those of the others, and is called the foramen of Monro. Extending backwards from the foramen of Monro, beneath the corpus fimbriatum or edge of the fornix, as far as the apex of the middle cornu of the lateral ventricle, is a narrow fissure which transmits the velum interpositum to the lateral ventricle. This fissure is called the fissure of Bichat, and is the ventricular termination of the great transverse fissure.

<small>Foramen of Monro Figs. 4-5.</small>

<small>Fissure of Bichat Figs. 4-5.</small>

The fornix is a triangular layer of longitudinal white fibres, with its apex directed forward and its base attached behind to the posterior extremity of the corpus callosum. Its anterior extremity divides into two rounded cords, one on each side, which descend in front of the thalamii optici to the base of the brain, and obliquely backward to the corpora albicantia in the interpeduncular space, and from these latter are tracts of white fibres passing outward to the internal capsules.

<small>Fornix. Fig. 4.</small>

The apex of the fornix is continued forward to the genu of the corpus callosum in the median line, by a thin membrane, the septum lucidum. This membrane fills up the interval between the fornix and the genu of

<small>Septum lucidum. Figs. 4-5-6 8 30.</small>

the corpus callosum dividing the lateral ventricles from each other. The fibres of the septum lucidum pass in a longitudinal direction from the fornix to the corpus callosum. The septum is composed of two layers, side by side, enclosing a vertical slit, the fifth ventricle. The posterior angles of the fornix are extended outward, backward, and downward along the broad groove behind the arch or genu of the internal capsule, to be continuous with the layer of white fibres which cover the floor of the middle and posterior cornua. These extensions are called its posterior pillars, each pillar apparently dividing to form the hippocampus major and minor of the same side. The edge of the fornix is called the corpus fimbriatum, and rests upon the thalamus opticus. It extends behind the posterior extremity of the thalamus downward along the internal margin of the floor of the middle cornu and terminates in the apex of the ungual convolution.

The choroid plexus is a capillary network lying upon the thalamus and attached to the edge of the velum interpositum, from which it receives the blood vessels that supply it. It extends from the foramen of Monro in front over the thalamus opticus into the middle cornu of the lateral ventricle. It lies parallel with the fissure of Bichat and the corpus fimbriatum to their lower extremities, which terminate at the apex of the middle cornu. The choroid plexuses of the lateral ventricles and those of the third ventricle are connected through the foramina of Monro.

The fornix forms the roof the third ventricle. If it is divided transversely and its parts reflected, its anterior pillars will be seen descending in front as two rounded cords; and upon the under surface of its posterior part, crossing in lateral direction, are several transverse striæ called the lyra.

Marginalia:
- Fifth ventricle. Figs. 4-8.
- Genu of the internal capsule. Figs. 4-7-8.
- Corpus fimbriatum and posterior pillars of fornix. Figs. 4 5.
- Ungual convolution. Fig. 17.
- Choroid plexus. Figs. 4 5.
- Foramen Monro. Fig. 4.
- Anterior pillars of the fornix. Figs. 5 6 26.
- Lyra. Figs. 5 6.

DISSECTION.

These latter are fibres of the corpus callosum crossing in a transverse direction beneath the attachment of the posterior margin of the fornix to that of the corpus callosum.

Beneath the fornix, and exposed by its removal, is a process of pia mater extending into the ventricles of the brain, and called the velum interpositum on account of its being interposed between the fornix and the third ventricle, which is beneath it. The velum interpositum is triangular in shape, corresponding to the under surface of the fornix, with which it is in contact. The velum contains numerous blood vessels which supply the choroid plexuses. Near its centre, side by side, are two large veins, the vena Galeni, which receive blood from the venæ corpora striata and convey it to the straight sinus, at the junction of the falx cerebri and tentorum cerebelli. Suspended along the middle line upon the under surface of the velum are two fringes, the choroid plexuses of the third ventricle, which are continuous with those of the lateral ventricles at the foramina Monro. The velum is derived from the pia mater covering the occipital and temporal lobes of the brain. It is admitted into the third ventricle beneath the posterior border of the corpus callosum, and passing forward under the fornix, beneath its lateral margins, it extends into the lateral ventricles through the fissures of Bichat. At the outer sides of the cruri cerebri, the pia mater passes directly from the under surface of the brain into the middle cornua of the lateral ventricles upon each side. This membrane embraces the pineal body, and the latter may be torn off by its careless removal. The parts exposed by its removal are, the cavity of the third ventricle, pineal body, optic lobes and the thalamus. These, together with the basal ganglia and internal capsule, will now be described together, in order that their relations may be more clearly

Velum interpositum. Fig. 5.

Situation in which the pia mater passes directly into the lateral ventricle from the surface of the brain. Figs. 5-17.

Parts exposed by the removal of the velum. Fig. 6.

understood. The removal of the cerebellum, as seen in Fig. 8, is advantageous to this dissection, as the parts above mentioned are then seen in connection with those below.

Dissection represented by Figs. 6-7.

In general, this dissection presents a median longitudinal depression, and two lateral masses, one on each side. In the middle line, between the lateral masses, is a longitudinal cleft, the third ventricle. The third ventricle is one and half an inch in length, extends vertically to the base of the brain,

Floor of third ventricle. Fig. 7-9.

where its floor is formed by the parts contained in the interpeduncular space, including the lamina cinerea, optic commissure and tuber cinereum. Communicating with the ventricle below is a canal leading to the pituitary body, the infundibulum. The pillars of the fornix descend in the anterior extremity of the third ventricle. By separating the anterior

Anterior commissure. Figs. 6 13- 14 21 22 23-14- 15*.*

pillars of the fornix the middle portion of the anterior commissure is exposed, and is seen crossing the anterior wall of the ventricle in front of them. The anterior commissure of the third ventricle is directed outward and backward on each side to the extremity of the inferior ramus of the internal capsule of the hemisphere, its extremities terminating in the temporal lobes. It is supposed to be the commissure of the centres of smell and taste. The internal surfaces of the thalami form the lateral walls of the third ventricle.

Third ventricle and its commissures. Figs 6 26.

Crossing its middle, uniting the central part of one thalamus with that of the other, is a broad gray commissure, the middle or gray commissure of the third ventricle. The ventricle is pointed behind, and is continuous with a canal which passes backwards beneath another transverse band of white fibres, the posterior commissure of the third ventricle. The canal just mentioned leads downwards and backwards, beneath the posterior commissure and the commissure of the corpora quadrigemina and valve of Vieussens, to communicate with

DISSECTION. 27

the fourth ventricle. This canal is called the aqueduct of Sylvius, or the iter e tertio ad quartem ventriculum, or, for short, the iter. Extending from the anterior pillars of the fornix forward to the genu of the corpus callosum, and separating the lateral ventricles, is the septum lucidum.

The iter. Figs. 29-27-32 33.

Behind the posterior extremity of the ventricle, and above its posterior commisure, is a small globular body somewhat elongated from before backwards, the pineal gland. Two delicate white tracts extend forward from the anterior extremity of the pineal gland, one on each side, upon the lateral margins of the third ventricle, they are the anterior pillars of the pineal gland and mark the division between the superior and internal surfaces of the thalamus. These tracts are lost in front of the anterior extremity of the thalamus, where they are in conjunction with the descending anterior pillars of the fornix. The anterior extremity of the pineal gland is attached to the upper surface of the posterior commissure of the third ventricle by a delicate membrane, whose lateral margins are thickened and called the posterior pillars of the pineal gland. The posterior margin and sides of the pineal gland are free, and it rests above, in front of, and between the anterior tubercles of the corpora quadrigemina.

Pineal gland and pillars. Figs. 6 8

The corpora quadrigemina or optic tubercles are four small bodies placed above and between the crura cerebri, and in the interval between the posterior extremities of the thalami optici. They are enclosed behind by a notch upon the anterior part of the cerebellum the incisura anterior. The corpora quadrigemina are in pairs, an anterior pair, the nates, and a posterior pair, the testes. The tubercles of each pair are placed one on either side of the median line and are connected by a commissure. Directed obliquely outward and forward from the nates and the testes, on either side, are two bands

Corpora quadrigemina. Figs. 6 8-9-10-27.

Incisura anterior. Figs. 6 27 28.

Brachia. Figs. 6 8-9 13-14.

of white fibres, the brachia, anterius and posterius, the former being connected with the nates, and the latter with the testes. These tracts are the external margins of the horizontal layers of fibres which pass from the corpora quadrigemina, beneath the posterior extremity of the thalamus, to the internal capsule. They are also connected with the posterior extremities of the optic tract and geniculate bodies, the brachium anterior with the corpus geniculatum externum and the brachium posteri with the corpus geniculatum internum. Behind the corpora quadrigimina, and covered by the cerebellum, are two broad and rounded cords, one on each side, which ascend from the hilus of the cerebellum and are the processes, or the superior peduncles of the cerebellum. These are united along their inner margins by a thin translucent membrane, which is depressed upon its upper surface, and is called the valve of

Processes and valve of Vieussens. Figs. 8 9-10-15 26 30 32.

Vieussens. The valve is pointed at its anterior extremity which projects upwards between the testes, and is elevated above the general surface. On each side of its anterior extremity are the apparent origins of the fourth nerves. The under surface of the valve and of the processes form the roof of the fourth ventricle. The superior vermiform process of the

Superior and inferior vermiform processes of cerebellum. Figs. 26-30 31.

cerebellum rests upon the valve of Vieussens and the inferior vermiform process is beneath it in the fourth ventricle. Upon the outer surface of each processus at its anterior part, crossing it obliquely in an upward direction, is a band of fibres parallel to those of the crusta from which it is divided by the sharp vertical furrow previously mentioned. This band of fibres terminates above in the testes and brachium posterior, and below it is continuous with the fundamental root zone of the spinal cord.

Corpus geniculatum internum. Figs. 7 8-9.

At the upper extremity of the depression between the crusta and the tract just mentioned is a tubercle, which is connected behind with the brachium posterius, and in front with the **lower** division of the optic tract, and is called the corpus geniculatum internum. External and above the corpus geniculatum internum, upon the outer division of the extremity of the optic tract, is an **elevation,** the corpus geniculatum **externum.** These bodies **are** apparently the outer terminations **of the brachia,** but, as before stated, the major part of the fibers of the **brachia passes beneath the** thalamus and optic tract to the internal capsule.

<small>Corpus geniculatum externum. Figs. 7-8 9.</small>

<small>Connection of the brachia with the internal capsule. Figs. 13 14.</small>

THE BASAL GANGLIA:—Upon each side of the third ventricle, resting upon **the upper extremities of** the peduncles or **cruri cerebri,** are two large **masses of gray** matter, one on each side, the basal ganglia of the cerebum. Each mass is composed of three large ganglia, two of which are **internal, and one external to a** broad and thick layer of **white fibres, the internal capsule.** The capsule is seen projecting above, from between the external and internal ganglia, as an elevated tract composed **of transverse** lamina or bundles of fibres. The fibres of the anterior **portion of either internal capsule** are directed forward **toward the anterior lobe of the cerebrum, the middle fibres upwards to the parietal region, and the posterior backwards and outwards into** the occipital and temporal lobes of the brain. The internal capsule **radiates like the pinna** of the ear, an interspace being left **opposite the anterior** perforated space **between the frontal and temporal lobes** of the hemisphere. **In consequence of this** disposition of the fibres of the capsule, **it encloses a conical concavity, open** externally, which is filled by a mass of gray **matter, the lenticular nucleus,** or external basal ganglion.

<small>Basal ganglia and their relation to the internal capsule. Figs. 6 7 8 9-10, 10* to 18*.</small>

<small>Distribution of the internal capsule. Figs. 7 9-13.</small>

<small>Interspace in capsule. Figs. 9 12, 18 to 20.</small>

<small>Lenticular nucleus enclosed as in a funnel. Figs. 7 9 12.</small>

At its posterior part, the internal capsule sweeps around **an** abrupt curve as it descends to **form the** roof of the middle cornu of the lateral ventricle. **The** reflected portion of the capsule is distributed to the temporal lobe and its abrupt deflexion behind is called its genu.

Genu of internal capsule. Figs. 7, 8.

THE INTERNAL BASAL GANGLIA :—The internal basal ganglia rest upon and are attached to the internal surface of the internal capsule. They are separated from each other by a depression, and a narrow band of white fibres, the tænia semicircularis. The superior internal ganglion, or caudate nucleus, rests upon the upper portion of the internal surface of the internal capsule, and projects into the floor of the lateral ventricle. Its anterior portion, or body, is large, and occupies the anterior cornu of the lateral ventricle. The posterior part or tail of the caudate nucleus gradually tapers from the body in front to form a narrow elevation which runs along the outer margin of the lateral ventricle, and, bending beneath the genu of the internal capsule, enters the roof of the middle cornu, at the end of which it terminates in a bulbous extremity.

Caudate nucleus. Figs. 4, 6, 7. 14 (N C)*

The tænia semicircularis is a narrow band of white fibres at the bottom of a depression which separates the caudate nucleus from the thalamus opticus. It extends along the whole length of the lower margin of the caudate nucleus. Beginning in front immediately anterior to the anterior pillars of the fornix as a broad surface, it extends backward as a narrow band which terminates below at the end of the middle cornu and is attached to the apex of the ungual convolution.

Tænia semicircularis. Figs. 6, 7.

The tænia semicircularis in the floor of the lateral ventricle has the appearance of a narrow band of white fibres, but, when the caudate body is removed, it is seen to be the margin of a tract of fibres which is derived from slips emerging from between the lamina of the internal capsule to increase its volume as it passes forward beneath the caudate nucleus.

Derivation of the tænia semicircularis. Left side Fig. 8.

DISSECTION. 31

THE THALAMUS OPTICUS:—The inferior internal ganglion, or thalamus opticus, in general outline is fusiform in shape. Its body is somewhat elongated, and a transverse section shows that it is prismatic. Its upper surface is divided **into two portions** by an oblique depression which begins in front behind the anterior pillar of the fornix, and extends backward and outward to **its** external margin at the genu of the internal capsule. The surface **above and external** to this depression forms a part of the floor of the **lateral ventricle.** The portion below **or** internal to the depression is **covered** by the fornix, and is continuous behind **and below** the **crus cerebri** with the outer division of the **optic tract.** The internal surface of the thalamus forms the lateral wall **of the third ventricle,** and in the middle of the ventricle, is united to the **thalamus of the** opposite side by **the middle or gray commissure of the** third ventricle. The inferior **surface of the thalamus looks** outwards and downwards and is attached to the **internal capsule to** which it gives off a great mass **of fibres,** before and behind, forming **the anterior and the posterior divisions of the internal capsule.** A **tract of longitudinal fibres,** called the fillet, lies external and **beneath the lower portion of the inferior surface of the thalamus.**

Upon closer **inspection the thalamus will be seen to** consist of several smaller **masses, more** or less similar in shape to that of the caudate nucleus. The anterior **extremities of** these masses, composing the thalamus, are large, **and their** posterior portions taper into tracts, which pass **backwards** around the genu of the **internal capsule. These bodies are,** from within outwards: first, a small **body situated at the side** of the posterior extremity of the third ventricle, between **the** pineal gland and the posterior extremity of the thalamus, **and** connected with the anterior pillar of the pineal gland,

Thalamus. Figs. 6-8 10.

Groove on it supper surface. Figs. 6 8.

Internal surface. Fig. 26.

Distribution of fibres from thalamus. Figs. 20-21.

Relation of the thalamus to the fillet. Figs. 10-13 21-22-23.

Masses comprising the Thalamus. Fig. 6.

1 The pison.

called the pison (pea). This body has no posterior extension. Second, a fusiform body, beginning in front by a pointed extremity above the middle commissure, and rapidly enlarging forms a prominence upon the posterior portion of the thalamus, after which it gradually contracts and is continuous with the external geniculate body and the superior division of the optic tract. This body is called the pulvinar and is probably the visual centre of the thalamus. Third, a prominence upon the anterior extremity of the thalamus, in front of the apex of the pulvinar, behind the anterior pillars of the fornix and above the foramen Monro, called the tuberculum of the thalamus. Extending backward and slightly outward from this tubercle is a tract of white fibres which follows the groove, before mentioned, on the upper surface of the thalamus as far as the genu of the internal capsule, among the fibres of which it is lost. The tuberculum thalami is supposed to be the basal centre of taste and smell. Fourth, external to the tuberculum and the tract just described, and more distinctly seen if the latter is removed, is the largest body composing the thalamus. It is broad in the centre, pointed at each extremity, fusiform in shape, reaching across the upper surface of the internal capsule, and separated from the caudate nucleus by the tænia semicircularis. This body is supposed to be the thalamic centre of common sensation. Fifth, underneath the central part of the thalamus, and imbedded in the tegmentum of the crus cerebri, is a round body, somewhat larger than a pea, called the subthalmic ganglion, or red nucleus. This body does not properly belong to the thalamus, but is closely associated with it. It is the termination of the anterior extremity of the processus, or superior peduncle of the cerebellum, and from it are distributed fibres which pass to the bodies

Marginal notes:
2. The pulvinar.
3. The tuberculum thalami.
4. The fusiformis.
5. Subthalmic ganglion or red nucleus of the tegmentum. Figs. 10 13 14-15-19°-20°.

composing the thalamus under which it lies. The **red nucleus** is in relation externally with the fillet, and internally with the continuation of the gray matter which surrounds the iter and lines the wall of the third ventricle.

THE EXTERNAL BASAL GANGLION:—The gray matter **external to** the internal capsule, and enclosed by its funnel **shaped concavity**, is the lenticular **ganglion**, so called **because its parts are disposed in lenticular layers. Its layers are superimposed and separated from each other by laminæ of white fibres, called capsules. A transverse section of the entire mass** of the lenticular **body, including the parietal and temporal portions of the internal capsule, is triangular. The apex of the triangle is** directed downward **and** inward, and its base is opposed to that of the insula. **It will** be seen that **the lenticular nucleus is** bounded **above and** below **by the internal capsule and is in** three layers or parts. The outer surface of each **layer is convex and composed mostly of gray matter, while the inner portion of each section is traversed by numerous white tracts for some distance from the septum or capsule. The three bodies which** comprise the **lenticular nucleus increase in size and thickness** from within outwards. The innermost is a conical wedge spreading the fibres of the internal capsule at the bottom of **its funnel** shaped concavity. The inner and **middle portions are together called the** globus pallidus, on account **of being much paler in color than the outermost, which is called the putamen. Externally the outer body or portion of the** lenticular nucleus is covered by a thick layer **of radiating white fibres which** diverge from below upwards.

The layer of diverging fibres external to **and** overlying the lenticular body, forming the external capsule, is thicker above than below, and is united with **the outer surface** of the internal capsule above and along the upper margin of the lenticular nucleus. The external capsule is covered upon its outer surface by a thin layer of gray matter which is thicker below than above, and is called the claustrum, (a wall,) external to which is the subinsular layer of white matter, and the insula. These parts therefore enclosed by the funnel shaped whorl of the fibres of the internal capsule are, from within outward, the three divisions of the lenticular nucleus, the external capsule, claustrum and the white and gray matter of the insula, the latter being enclosed in the Y shaped expansion of the bottom of the fissure of Sylvius.

Claustrum. Figs. 10ᵃ to 18ᵃ ic 1,

Parts enclosed in the internal capsule from within outward. Figs. 5-15ᵇ-10ᵇ, between (IC and IC).

The internal capsule will be studied in detail subsequent to the description of those parts below which enter into its formation. It is sufficient at this time to remember its shape, its relations to the basal ganglia and that fibres from these ganglia contribute to swell its volume.

Because of the funnel shape of the internal capsule, a horizontal section of the brain, as generally represented in diagram, shows an anterior and a posterior arm, and locates the motor portion of the capsule behind the knee in the anterior part of the posterior arm.

The hemispherical ganglion and lateral dissections of it will be described after its internal anatomy has been traced in a longitudinal direction. The crura cerebri or peduncles of the cerebrum, as before stated, consist of two portions, the crusta and the tegmentum. The crusta is a broad band of fibres upon the outer and anterior part of either crus cerebri. The tegmentum is all that portion of the peduncle behind and internal to the crusta. The crusta originates entirely from the pons Varolii, but the tegmentum is connected with the cerebellum and with the posterior layer of the pons which includes the fillet and fasciculi teretes, and

Crusta and tegmentum. Fig. 32.

through these with the spinal cord. The connections and the relations of the various tracts which compose the peduncle of the cerebrum will be more clearly understood after the structure of the cerebellum and the pons Varolii have been studied. We will again return to the anatomy of the cerebrum in order to trace the parts of the peduncle to their connections with those of the hemisphere.

CEREBELLUM.

Cerebellum. Figs. 6 7 27 to 31 24* to 26*

It will not assist us to the understanding of the architecture of the central nervous system to enter into a minute description of the external surface and shape of the cerebellum. Its external appearance and the divisions of its surface will be readily understood by a study of the plates in this work. It is sufficient to say that it is composed of a middle and two lateral lobes. The upper portion of the middle lobe is called the superior, and the lower part the inferior vermiform process of the cerebellum. The superior vermiform process rests upon the upper surface of the valve of Vieussens, and the inferior vermiform process is beneath the valve, and within the fourth ventricle. The lateral lobes project backwards and outwards, and are large masses separated in front by a notch, the incisura anterior, and behind by a deep and narrow fissure, the incisura posterior. In front of the lower surfaces of the lateral lobes, beneath the lower margin of the pons, are two small lobes, one on each side of the medulla oblongata, called the flocculi.

Superior and inferior vermiform processes. Figs. 26-27 to 31.

Incisura anterior and posterior. Figs. 6 29.

Flocculi. Figs. 7-28.

The cerebellum is connected with the cerebrum by the processes which are its superior peduncles; with the pons in front by its middle peduncles; and with the spinal cord below by two rounded cordlike bodies, the restiform bodies, or its inferior peduncles.

Peduncle of the cerebellum. Figs. 8 to 15½.

A section of the middle lobe of the cerebellum in the median line from before backward, exposes the fourth ventricle. The surface of the section presents a beautiful arborescent distribution of white matter radiating from the end of a longitudinal section of the valve of Vieussens the rami from which divide into small branches as they approach the surface. These subdivide into minute stems of white **fibres terminating** in small leaflets which line the sides of the **deep fissures that project** inward from the surface of the lobe. The appearance presented by this section is called the arbor vitæ.

Arbor vitæ. Figs. 29 30.

A section across a lateral lobe, outward and backward, shows within, a large mass of white matter, an arborescent border, and imbedded in the anterior part of the white substance a convoluted layer of gray matter surrounding a portion of white substance of irregular shape. This body is called the corpus dentatum of the cerebellum and is its sensory or internal basal ganglion.

Section of a lateral lobe of the cerebellum. Fig. 31.

Corpus dentatum. Figs. 8 to 15½.

In a properly hardened cerebellum, if the sides of its transverse fissure are separated, and the upper half gently torn from the lower and broken off in front, it will be observed that the surface of either lobe will separate behind and in front over the dentate body, that the fractured fibres are external to this body, and are those of the middle peduncle of the cerebellum.

Horizontal division of the cerebellum. Fig. 15½.

The upper surface of the dentate body is elevated above the remaining portion of the surface thus exposed upon the inferior segment, or lower portion of the cerebellum. This surface presents numerous lines radiating outward from the corpus dentatum, and others directed inward and backward, from the broken middle peduncle of the cerebellum. The fibres from these separate sources commingle to form the corona radiata of the cerebellum.

Dentate body. Figs. 8 to 15½.

Corona radiata of of the cerebellum Fig 15½.

It will be observed that the processus, or superior peduncle of the cerebellum, of either side passes directly downward and backward into the centre of the corpus dentatum, entering its hilus upon the under surface. The lower extremity of the processus, on its outer side, is overlapped by a tract of radiating fibres which spring upward from the interval between the middle and superior peduncles of the cerebellum as they approach each other. This tract emerging from between the superior and middle peduncles, is the upper extremity of the restiform body, or inferior peduncle of the cerebellum, and is distributed to the upper surface of the corpus dentatum.

Termination of the restiform body in the corpus dentatum. Fig. 15½.

The attention is directed to the fact, that the corpus dentatum is connected with the superior and the inferior peduncles of the cerebellum, directly associating this body with the cerebrum above, the spinal cord below, and the cerebellum behind. The middle peduncle is composed of two distinct layers of fibres, differing in function, connected directly with the gray matter of the surface of the cerebellum and having no connection with the corpus dentatum.

Corpus dentatum connected with the superior and inferior peduncles. Figs. 8 9-15½.

Thus it is apparent that there is an analogy between the structure of the cerebellum and that of the cerebrum and its application at this time will assist us to understand those parts of the pons Varolii and medulla oblongata which are associated in the comparison.

Comparison of the parts of the cerebellum to those of the cerebrum.

The sheet of gray matter which covers the surface of the cerebellum, and dips into its fissures, is analagous to that which covers the hemispheres of the cerebrum, and may therefore be called the hemispherical ganglia of the cerebellum. The enclosed mass of white matter, composed of the interlacing fibres of the middle peduncle and the corpus dentatum, constitute the corona radiata of the cerebellum The corpora dentata, associated with the sensory tracts of the

Hemispherical ganglion of the cerebellum.

Corona radiata of the cerebellum.

cerebro spinal axis, correspond to the internal basal ganglia of the cerebrum, and, as we shall see subsequently, the external basal ganglia of the cerebellum are displaced, and are located in the pons Varolii. The superficial layer of the pons unites the opposite hemispheres of the cerebellum, and is the analogue of the corpus callosum of the cerebrum. The deep layers of fibres of the middle peduncles terminate in the gray matter of the pons Varolii, and correspond to the motor tracts of the internal capsules of the cerebrum, and the crustæ of the crura cerebri. The superior and inferior peduncles of the cerebellum are interrupted in their passage by the corpora dentata, thus they coincide with the tegmentum of the cruræ cerebri. The direct connection of the cerebellum with the spinal cord, by means of the direct cerebellar tract and the arciform fibres of the medulla oblongata, may correspond with the fillet of the cerebrum, completing the analogy between these large masses of cerebro spinal axis. Taken together, the superior peduncle, the internal portion of the middle peduncle, and the inferior peduncle of the cerebellum, strikingly correspond to the several elements which enter into the formation of the internal capsule of the cerebrum.

By the use of the above analogy the study of the connecting parts is made easy.

Corpora dentata, the internal basal ganglia.

Superficial fibres of the pons, the corpus callosum of cerebellum.

Deep layer, the motor tract of its internal capsule.

Superior and inferior peduncles correspond to the tegmentum.

Arciform fibres compared with the fillet.

THE PONS VAROLII.

Description of the pons Varolii.

The pons Varolii is a broad and thick band of transverse fibres crossing in front of the upper continuation of the spinal cord and the medulla oblongata, interrupting the anterior median fissure, and connecting the opposite hemispheres of the cerebellum. It is about one inch and a half broad from above downward, and nearly two inches in a lateral direction. It is contracted upon each side as it approaches the cerebellum, where it forms the middle peduncles of that body. Its general shape is oval, and its anterior surface is convex in every direction. A shallow depression crosses its surface, from above downward in the median line, called the basilar groove, which lodges the basilar artery. This portion of the pons is perforated by numerous openings for the passage of blood vessels. The upper margin of the pons is bounded in the middle by the posterior perforated space and laterally by the fibres of the crustæ. The third nerves arise from the inner sides of the crustæ above this margin. Its lower margin is bounded in the middle line by the anterior median fissure and raphe of the medulla oblongata, and successively on each side of the fissure, from within outwards, by the anterior pyramid, olivary body, lateral tract, and the restiform body. Behind the pons Varolii is the floor of the fourth ventricle. The flocculi rest upon the outer extremities of its lower border in front of the restiform bodies. The sixth nerve emerges from between the pyramid and the olive, the seventh and eighth from internal to the flocculus and resti-

Fourth ventricle. Figs. 8-26.

Flocculi. Figs. 7, 28.

Fifth, sixth, seventh, eighth nerves. Figs. 9-10.

THE PONS VAROLII.

form body, and each of these nerves is directed **forward** beneath the lower **border** of the pons. The fifth **nerve** emerges from **the middle** of the anterior surface **of the middle peduncle of the cerebellum.**

The **lower fibres** of the superficial layer of the pons, which form the **lower third of its** anterior surface are transverse in direction, and **are overlapped** at their outer extremities by those from above which **arch downwards near their termination.** The upper transverse fibres are arched upward in the middle and terminate in the lower lobes **of the cerebellum on** each side. The uppermost fibres of **the pons are seen to proceed** from the **posterior perforated space, and winding close to the crus cerebri on each side are directed backwards to enter the cerebellum** above the processes. These latter fibres do not properly belong to the pons Varolii, but are **derived from the floor of the fourth** ventricle, and **passing forward through the raphe connect the** gray **matter of the fourth ventricle** with the cerebellum. **They are similar in function to the restiform body of the medulla** oblongata **and are distributed to the corpus dentatum in** conjunction **with the inferior peduncle of the cerebellum.** Other fibres, derived from the **floor of the fourth ventricle, enter the raphe and pass downwards** beneath the lower margin of **the pons. These emerge** in front from the **anterior median fissure of the medulla** oblongata and winding over **the pyramid and olive, in conjunction** with the arciform **fibres, join** the **restiform bodies to** terminate in **the corpus dentatum of the cerebellum.** Doubtless, **also, there are associated with these,** other fibres that correspond **to the direct cerebellar tract of the spinal cord, and which will be better understood after this tract has been described.**

See pons. Figs. 7-9 to 14.

The uppermost fibres of the pons Figs. 9-10.

Inferior restiform fibres of the floor of the fourth ventricle. Figs. 9 10

STRUCTURE OF THE PONS VAROLII:—The several layers of which the pons Varolii is composed, will be more easily understood by dividing it in the median line from before backward, into lateral halves. After the cerebellum has been removed, if the pons Varolii is torn in a longitudinal direction along the floor of the fourth ventricle in its posterior median fissure, it will separate into striated surfaces to the depth of about a quarter of an inch in the middle of the ventricle, and above the pons, between the cruri cerebri, it will divide as far forward as the posterior perforated space, and below to the anterior median fissure of the medulla oblongata. The adjacent surfaces, thus separated, will be seen to form each a plaino-concave surface, concave in front and straight upon its posterior margin along the floor of the fourth ventricle. If the remaining portion of the pons is now divided in the median line by the knife, the pons will be separated into lateral halves. The surface, presented by a section of the anterior portion of the pons, is lenticular in shape and covers the anterior two-thirds of the whole surface thus exposed.

It will be observed that in a division of the pons Varolii, transverse to the fibres on its anterior surface, and longitudinal to the cerebro spinal axis, the divided surface presents two distinct portions for examination; an oval portion in front which is connected with the anterior pyramid of the medulla oblongata, and with the crusta of the crus cerebri; and behind, a plano-concave surface of radiating fibres, the raphe, which separates the lateral halves of the crura cerebri, the posterior part of the pons, and the medulla oblongata. The raphe is composed of anterio-posterior fibres, and is an important structure, because, within it are the decussations of the nerves of the medulla oblongata and the commissures of the nerve centres situated in the floor of the fourth ventricle.

THE PONS VAROLII.

Upon each side of the posterior margin of the raphe, upon the floor of the fourth ventricle, and that of the iter, are two rounded cords, the fasciculi teretes. These are broad in the ventricle below, narrow above in the floor of the iter, and are continuous in front with the gray matter upon the lateral walls of the third ventricle. These columns attend the posterior margin of the raphe as far as the floor of the third ventricle. The superior margin of the raphe curves abruptly in the floor of the third ventricle forward to the base of the brain in front of the corpora albicantia.

Fasciculi teretes. Figs. 15 35.

Downward curve of the upper border of the raphe in the floor of the third ventricle. Fig. 26.

In that part of the raphe which separates the crura cerebri is an oval opening about the size a split pea, around which the fibres of the raphe separate to form its margins. This opening in the raphe is filled by the extremity of a tract of broken fibres which are those of the decussation of the processes, or superior peduncles of the cerebellum. The fibres of the processes pass beneath the fasciculi teretes, and decussate within this opening in the raphe. The processus of the opposite side crosses the raphe at this point to reach the red nucleus which is situated external to the raphe above the point of decussation. (The letters RN Figure 26, are placed over the situation of the red nucleus which is buried within the tegmentum of the crus cerebri.) The opening in the raphe, just described, is above the superior margin of the pons, and in the middle of that portion of the raphe which separates the crura cerebri from each other.

Oval opening in the raphe. Fig. 26 (PT).

Red nucleus. Figs. 10 13 to 15½-19°-29° (RN).

By the removal of the raphe there will be seen, behind the pons, the margin of a longitudinal tract of white fibres extending from the medulla oblongata to the crus cerebri above. This tract is the internal lateral margin of the lemniscus or fillet, which is separated from its fellow by the raphe. This tract is crescent in shape, with its concavity

Internal margin of the lemniscus or fillet. Fig. 26 (PT).

applied to the posterior convexity of the lenticular section of the pons before described. Behind the fillet is a layer of gray matter which underlies the floor of the fourth ventricle, and constitutes the fasciculа teretes.

Fasciculi teretes. Figs. 15, 35.

The pons, therefore, consists of two distinct portions; an anterior portion composed of transverse and longitudinal fibres, the deep layer of which is mixed with gray matter, a section of which is lenticular in shape; and a posterior portion, plano-concave in shape, and composed of two layers, which is the continuation of the spinal cord and medulla oblongata, except the anterior pyramids in front, and the restiform bodies behind. These two portions of the pons Varolii are easily separated from each other by dividing the anterior pyramid and gently tearing the transverse fibres of the pons from those behind which have a longitudinal direction. These portions of the pons are distinct in function, that in front being commissural containing the great commissure of the cerebellum and the motor tract, and that behind belonging to the sensory tract and the column of gray matter called the fascicula teretes, which is a continuation of the gray column of the spinal cord.

Recapitulation of the structure of the pons Varolii.

These parts may be more readily understood by a dissection from before backwards. The anterior division of the pons is composed of two layers, and the posterior division also of two layers, which will be described in succession. The superficial layer of the anterior division of the pons has before been described as consisting of three parts, an inferior portion which has a transverse direction, a middle portion which is transverse on each side of the basilar groove and laterally arches downward to overlap the extremities of the inferior fibres, and an upper portion which emerges from the posterior perforated space and passing outwards and backwards around

Anterio-posterior dissection of the pons. Figs. 11 to 14.

the crura cerebri to terminate in the cerebellum along with the restiform bodies. The superficial layer of the pons Varolii is the great transverse commissure of the cerebellum, excepting its superior fibres, which are the restiform fibres from the floor of the fourth ventricle, as before stated. If this layer of fibres, which is the commissural layer, be divided in an anterio-posterior direction, it will be seen to average one sixth of an inch in thickness. Beneath this layer is a network of transverse and longitudinal fibres interspersed with gray matter, and is called the formatio reticularis of the pons Varolii. The network is composed of the deep transverse fibres of the middle peduncle of the cerebellum, interlacing with fibres which descend from the crustæ of the crura cerebri. The fibres of a crusta are disposed in laminæ, similar to those of the internal capsule, and are placed one behind the other in the crus, but as they descend into the pons they are separated from each other in order that the transverse fibres of the deep layer of the pons, also disposed in laminæ, may be inserted between those of the crusta as they descend. The vesicular matter of the pons is interspersed throughout this formatio reticularis, but is more concentrated in the middle of the posterior portion of this part of the pons. The gray matter of the pons is called its nucleus, and is continuous above with the locus niger or vesicular tract of the crus cerebri.

The fibres derived from the anterior pyramids of the medulla oblongata pass upward with the longitudinal fibres of the pons to the crusta and internal capsule. The reticulated layer is the motor portion of the pons Varolii. It consists of the crusta, the fibres of the anterior pyramid, the deep layer of the transverse fibres of the middle peduncle, together with the gray matter, or the ganglion of the pons Varolii. The

Superficial layer of the anterior portion of the pons is the great commissure of cerebellum. Figs 9 31-35.

Deep layer of the anterior portion of the pons or formatio reticularis. Figs. 9-34 35.

Nucleus of the pons. Figs. 34-35.

Locus niger. Figs. 10 to 17 32 33. 20 21* (SN).*

ganglion of the pons is the motor basal ganglion of the cerebellum, and is the analogue of the lenticular ganglion of the cerebrum.

The ganglion the of pons is the external basal ganglion of the cerebellum.

The posterior division of the pons Varolii, or that portion of it which is a continuation of the spinal cord, can be better studied by the removal of the part which has just been described. If the anterior pyramids and the lateral attachments of the pons to the cerebellum are divided, and the anterior portion of the pons is gently lifted from its bed, it will separate from a broad band of longitudinal fibres, extending from the medulla oblongata to the crura cerebri. By a division of the crusta above, the whole can be removed in a mass, leaving a broad excavation flat from side to side and concave from above downwards. This surface is divided into lateral halves by an elevated ridge of torn fibres which extends longitudinally from the anterior median fissure of the medulla oblongata to the center of the interpeduncular space above. The central ridge is composed of the ruptured fibres of the raphe. The surface exposed presents the appearance of a broad transverse groove, and lodges the anterior portion of the pons Varolii. Upon each side of the anterior median fissure, along the lower border of the groove, are the divided ends of the anterior pyramids which project upwards as the extremities of rounded cords. Upon the outer side of these are the upper extremities of the olivary bodies, external to which are the lateral tracts of the medulla oblongata, and more externally the arched fibres of the restiform bodies. Emerging from beneath the divided extremities of the anterior pyramids, and from between the olivary bodies and the raphe, are two tracts, one on either side, which ascend in an outward direction, obliquely to the upper and outer margins of this surface where they become superficial. They overlap the anterior part of

Longitudinal fibres of the Varolii. Figs. 11 to 14.

Elevated ridge dividing the lemniscus or fillet longitudinally in the middle line. See raphe. Figs. 11 to 14.

Anterior pyramids. See medulla oblongata. Figs. 9-11 to 14.

THE PONS VAROLII.

the processes and terminate in the testes and brachia posterius. These tracts are the continuation of the fundamental root zones of the spinal cord. They have been called the anterior layer of the lemniscus, or fillet, but are distinct **tracts of fibres** which connect the corpora quadragemina with the anterior horns of the gray matter of the spinal cord, and do not belong to the same system of fibres as the fillet. Beneath the layer just described, is a longitudinal layer of fibres derived from the olivary bodies of the medulla oblongata, and is called the olivary fasciculus. Under the olivary fasciculus is a similar layer which is continuous below with the lateral columns of the medulla oblongata, and the anterior root zones of the spinal cord. Behind this layer is one which is derived from the gray matter of the fasciculi teretes and which joins the fillet above and from the inner side of the restiform body. The three layers last described constitute the fillet proper, but it will be convenient in ordinary to say, that these four layers unite in forming the longitudinal layer of this portion of the pons Varolii. The outer margin of the fillet forms the internal boundary of a triangle, which is bounded above by the processus, and behind by the restiform body, and is called the restiform triangle.

The vesicular layer of the posterior portion of the pons Varolii underlies the floor of the fourth ventricle, and being elevated upon its posterior surface is called the fasciculi teretes. This layer is continuous below with the gray columns of the spinal cord, which, in the floor of the fourth ventricle, are separated behind, so that the posterior cornua are external upon the sides, and the anterior cornua lie together along the median fissure of the ventricle. On account of this disposition of the anterior and posterior cornua, the centres of the motor nerves of the medulla oblongata are

Fundamental root zone. Figs. 11-12-26.

Olivary fasciculus Fig. 12.

Anterior root zone or lateral tract of the medulla. Figs. 7-10-11-12-13.

Fillet fibres from the floor of the fourth ventricle. Figs. 11-12-13.

Restiform triangle Figs. 9 to 13.

Fasciculi teretes. Figs. 10-15.

placed along the median line of the ventricle, and those of the sensory nerves upon its sides.

The fourth ventricle. Figs. 8 10-15-26.

THE FOURTH VENTRICLE:—The fourth ventricle is situated behind the pons Varolii and the upper half of the medulla oblongata. It is bounded below, on each side, by the restiform bodies, and above by the inferior margins of the processes.

Roof. Figs. 8-26.

Its roof is formed by the processes on each side, and by the valve of Vieussens in the middle. It is about one inch and a half in length from above downwards, and about three-quarters of an inch in width at its widest part, which is between the upper portion of the restiform bodies, its depth from above downwards, from the valve of Vieussens to the median fissure, is about five-eighths of an inch. Its floor is lozenge shaped and pointed at either extremity, its upper extremity is continued by the iter to communicate with the third ventricle. Its lower extremity is continuous with the central canal and the posterior median fissure of the spinal cord. The median

Dimensions.

Floor of the ventricle. Figs. 8 10 15 26.

fissure of the fourth ventricle divides its floor into lateral halves, and is the posterior boundary of the raphe. The restiform bodies are elevated on each side above the level of the floor of the ventricle, and their gradual approach and junction at the lower extremity of the ventricle resembles a pen in appearance, hence the lower extremity of the ventricle is called the calamus scriptorius.

Calamus scriptorius. Figs. 8-10.

Clavate nucleii. Figs. 8 10.

Below are two slight elevations or grey tubercles, one upon each side of the apex of the ventricle, the clavate nucleii, which are the terminations of the posterior pyramids of the medulla oblongata. Directed upwards and outwards from these tubercles, along the floor of the fourth ventricle internal to the restiform body on either side, is a tract of white fibres which passes through the restiform triangle, to join the fillet. This tract forms the posterior layer of

Origin of the deep layer of the fillet in the fourth ventricle. Figs. 8 10 to 14.

the fillet, and a portion of it turns outward with the restiform body to the cerebellum. It also receives the anterior root of the eighth nerve as this root turns downward from the restiform triangle into the fourth ventricle to reach the auditory nucleus, which is situated beneath the floor of the ventricle at the side of calamus scriptorius.

Anterior root of the eighth nerve. Figs. 8-9.

At the lower part of the restiform body, external to the ventricle, and above the clavate nucleii, is a broad elevation covered by white fibres which marks the situation of the restiform nucleus. The restiform nucleus is the superior termination of the posterior column of the medulla oblongata and spinal cord, and it is the origin of a large portion of the fibres which constitute the restiform body.

Restiform nucleus. Figs. 8-10-36.

The floor of the fourth ventricle is gray in color, some parts of it presenting bluish elevations which are the centres of nerves, and are called loci cerulii. If the sides of its median fissure are separated, the torn surfaces present a striated appearance of antero-posterior oblique fibres, which intermingle across the median line. The fibres thus separated compose the raphe before described.

Loci cerulii.

Crossing the surface of the floor of the fourth ventricle, about its middle, are several striæ, which are the striæ transversæ. The inferior striæ are derived externally from the posterior root of the eighth nerve (auditory), which pass inward beneath and across the flexion of the restiform body to the floor of the fourth ventricle, to enter the raphe, in the median line, among the fibres of which they are lost. The superior striæ transversæ commence above the arch of the restiform bodies upon the outer margin of the floor of the ventricle, and are directed downward and inward, obliquely from the restiform triangle to the raphe.

Striæ transversæ. Figs. 8-10.

Posterior root of auditory nerve. Figs. 8-10

Root of the facial nerve on the floor of the fourth ventricle. Figs. 8-10.

They may be traced as far forward, and below, as the inferior margin of the pons Varolii, at the anterior median fissure of the medulla. These fibres are continuous outwardly on the same side with the seventh nerve (facial), and, internally, at the anterior median fissure, with the anterior pyramid of the opposite side.

Cavity of the fourth ventricle. Figs. 8-10-26-25* to 29*.

The roof of the fourth ventricle is arched from side to side, and diminishing in size gradually towards its anterior extremity encloses a conical shaped cavitiy which is contracted above to be continuous with the iter. This cavity lodges the inferior vermiform process of the middle lobe of the cerebellum, the superior vermiform process resting upon the upper surface of the parts forming the roof of the fourth ventricle.

The parts which form the roof the fourth ventricle, viz.: the processes and the valve of Vieusseus, which enter the upper part of the hilus of the cerebellum are connected behind with the dentate body. In front they assist in the formation of the peduncles of the cerebrum.

Recapitulation of the parts which enter into the formation of the pons Varolii.

To recapitulate: The pons Varolii is divided into two portions, au anterior or cerebellar, and a posterior or spinal. These portions include from before backwards: first, a transverse layer which is the commissure of the cerebellum; second, a reticulated and vesicular layer, which is the motor tract; third, a longitudinal layer, the lemniscus or fillet; fourth, a vescicular layer or the fasciculi teretes; and fifth, the restiform bodies or inferior peduncles of the cerebellum, which will be described with the medulla oblongata in connection with the tracts from which they are derived. The relation of these layers to each other should be recollected, in order to trace their distribution above in the formation of the crus cerebri and internal capsule, as well as their connections below with the medulla oblongata and spinal cord.

THE MEDULLA OBLONGATA AND SPINAL CORD.

These parts of the cerebro spinal axis are so intimately associated in general structure, and in the continuity of the tracts of grey and white columns of which they are constituted, that it will be convenient to describe them together. The medulla oblongata is properly the upper portion of the spinal cord, in which there is a rearrangement of its columns and the addition of important nerve centers. The spinal cord possesses the distinction, that in its structure the vesicular or grey matter is enclosed within the white matter, while in that of the brain the gray matter is external. *Medulla and spinal cord. Figs. 7 to 15.*

The cord including the medulla is about eighteen inches in length, extending from the lower border of the pons Varolii downward in the spinal canal to about the level of the lower border of the first lumbar vertebra. It terminates in a slender filament, the filum terminale, which contains a small amount of gray matter and ends at the sacral canal.

The spinal cord is rounded in transverse section, is divided into lateral halves by an anterior and posterior median fissure, the former extending upward to the lower border of the pons Varolii, and the latter to the separation of the posterior columns of the cord which form the sides of the fourth ventricle. About an inch and a half below the inferior border of the pons the anterior median fissure is partially obstructed by a tract of decussating fibres, which are those of the anterior pyramids of the medulla oblongata. The lower extremity of the decussation of the pyramids is the line of division *Transverse section Figs. 36 37 38.*

Decussation. Figs 11 37.

between the spinal cord proper which is the lower portion, and the medulla oblongata above. At the point of decussation there is an interruption in the grey elements of the spinal cord, produced by tracts of white fibres, which are directed forward from the deep fibres of the lateral columns of the cord, to become superficial in front of the medulla oblongata. The lateral halves of the cord are united in the middle by a commissure, which is white in front and grey behind. In the middle of the grey commissure is a minute canal, the ventricle of the spinal cord, which extends through its length and opens into the fourth ventricle at the apex of the calamus scriptorius. The fourth ventricle is a continuation of the central canal of the spinal cord, and is formed by a defect of the posterior commissure of the cord and a separation of its posterior columns into the restiform bodies. The fourth ventricle begins in the middle of the posterior surface of the medulla oblongata.

Commissure. Fig. 38.

The grey substance of the spinal cord is arranged in two columns, one on each side, united by a grey commissure which contains the central canal. The anterior portion of each column is called its anterior horn and is the motor vesicular column of the cord; the posterior projection or column of grey matter is called the posterior grey horn, or sensory arm. At the junction of the posterior cornu with the grey commissure is a slender vesicular column, called the vesicular column of Clarke. This column is probably the cerebellar element of the spinal cord, since it is connected directly with the cerebellum by the direct cerebellar tract, and degenerates along with it in the disease known as cerebello spinal ataxia.

Anterior and posterior horns of grey matter. Fig. 38.

Vesicular column of Clarke. Fig. 38.

The anterior vesicular columns of the spinal cord are disturbed or interrupted by the decussation of the anterior pyramids of the medulla about an inch and a half below the pons Varolii. As the vesicular columns of the cord approach the fourth ventricle the posterior cornua, or posterior vesicular columns, gradually diverge, and the anterior cornua approach each other toward the median line. In the floor of the fourth ventricle the grey columns are spread into a grey sheet, which forms the fasciculi teretes of the floor of the ventricle; the anterior columns lying along the median fissure, and the posterior along the sides of the floor of the ventricle.

The sensory roots of the spinal nerves terminate in the posterior vesicular columns of the spinal cord, and the sensory cranial nerves in masses of vesicular matter in the outer part of the floor of the fourth ventricle. The motor roots of the spinal nerves are connected with the anterior vesicular columns of the cord, and the motor cranial nerves with the motor centers along the middle of the floor of the ventricle.

Connections of the spinal nerves with the grey horns. Fig. 38.

The roots of the spinal nerves emerge from the sides of the spinal cord; the anterior roots in a line about one-third of the distance laterally on each side of the anterior median fissure, and the posterior roots also about one-third of the distance outward upon each side of the posterior median fissure, dividing the spinal cord on each side into three columns, anterior, lateral and posterior. Each column of the cord is subdivided into tracts which can be traced through the medulla to join with the several divisions of the pons Varolii described, and with the restiform bodies or inferior peduncles of the cerebellum.

Anterior, lateral and posterior columns of the cord.

The spinal cord is variable in size in different localities, due to the difference in the volume of the spinal nerves received by it at certain points. The lumbar enlargement is near the lower end of the cord and receives the nerves from the lower extremities. The cervical enlargement extends

Description of the spinal cord.

from the third cervical to the first dorsal vertebra, receives the nerves from the upper extremity and is somewhat larger than that of the lumbar region. The medulla oblongata is the uppermost enlargement of the spinal cord, it lies within the cranial cavity and is sometimes called the bulb. This part of the spinal cord contains the centers which control the functions of respiration and circulation, and is therefore one limb of the tripod which immediately sustains life. It also coordinates the muscles that perform the acts of deglutition and speech, besides other important functions. Symptoms indicating defection of this part of the central nervous system are of grave import, and are spoken of as bulbar symptoms, bulbar paralysis, etc. Certain diseases originate from functional or organic disturbance of the medulla, therefore a familiarity with its anatomy is fundamental to the diagnosis of these diseases, and also assists in the early recognition of those symptoms which denote the approach of death. The relations of its tracts to the several nerves of distinct function provide certain combinations of symptoms by which the anatomist can determine the exact situation and extent of a lesion.

<small>Nerves of the medulla. Figs. 7 & 9-10</small>

In order that the relations and the continuity of the columns and tracts of the spinal cord with those of the medulla oblongata, and the tracts of the latter with the divisions of the pons Varolii and with the cerebellum, may be more clearly understood, we will locate and name, first, the tracts of the spinal cord, then those of the medulla, and afterward associate them with each other, and with the divisions of the pons which have before been described.

The spinal cord proper, as stated, is that which extends below the decussation of the anterior pyramids of the medulla oblongata. It is uniform in structure, size and appearance except the enlargements of its cervical and lumbar regions, and that its fascicular tracts, or the white substance diminishes in quantity from above downward.

The grey columns and the central canal, or ventricle of the cord, have been described; the former as being continuous with the grey matter in the floor of the fourth ventricle, and the latter as a part of the ventricular system which extends through the whole length of the cerebro spinal axis.

It remains to trace the tracts which form the white columns of the cord and to describe their relations.

Tracts of the spinal cord. Figs. 7 to 15 38.

The relation of these tracts will be more easily understood by an examination of a transverse section of the spinal cord as represented in figure 38, which exhibits the division of the cord into three columns by the anterior and posterior roots of the spinal nerves.

Each of the three columns of the cord are divided into tracts, the anterior and the posterior each, into two, and the lateral into four. Each of these tracts are endowed with a specific function, and are arranged in bundles which can be separated in part by careful dissection, traced by the degeneration observed as the result of disease, and also by the observation of the development of the embryo.

Upon either side of the anterior median fissure is a small column which descends directly from the anterior pyramid of the medulla, called the direct pyramidal tract, or column of Türck. The remaining portion of the anterior column of the cord, which is much the larger part, is called the fundamental root zone, because it is among the first elements to develope in the spinal cord.

Direct pyramidal or column of Türck.

Fundamental root zone.

On either side of the posterior median fissure is another small column, called the posterior median column, or column of Gall, which is supposed to be the respiratory tract of the cord. The remaining portion of the posterior column of the

Posterior median or column of Gall

cord, which is much larger than the preceding, is called the posterior root zone, the cuneate fasciculus, or the column of Burdach. It is the sensory column of the spinal cord, disease of which disorders muscular co-ordination, as in the case of progressive locomotor ataxia.

The lateral column of the spinal cord is constituted of four fasciculi or tracts, and is included between the lines of emergence of the anterior and posterior roots of the spinal nerves. These tracts are: first—a ribbon like tract situated directly in front of the extremity of the posterior horn of grey matter of either side, and occupying the surface of the lateral column for about half its extent. This tract is connected in the spinal cord with the grey column of Clarke, and is continued above upon the lateral surface of the medulla and restiform body to the cerebellum. It suffers degeneration in the disease known as cerebello spinal ataxia, which runs a rapid course, affecting principally the nutritive functions of the body. Second—beneath the direct cerebellar tract, and anterior to the posterior horn of the grey matter, is a rounded fasciculus called the crossed pyramidal tract. This tract is connected with the vesicular matter of the posterior horn of the cord, and above with the anterior pyramid of the opposite side of the medulla oblongata. It decussates with the crossed pyramidal tract of the opposite half of the spinal cord in the anterior median fissure, at the lower extremity of the medulla. Beyond the decussation it unites with the opposite direct pyramidal tract to form the anterior pyramid of the opposite side of the medulla. The crossed pyramidal tract is the motor column of the cord, and possibly the office of the direct pyramidal tract is to co-ordinate the muscles of one side of the body with those of the other. Third— between the columns just described, and the anterior cornu

THE MEDULLA OBLONGATA AND SPINAL CORD. 57

of grey matter and the anterior roots of the spinal nerves, is a large column of fibres forming the anterior portion of the lateral column of the cord. It is called the anterior root zone, and in the spinal cord is supposed to contain longitudinal fibres that connect the segments of the cord together, and also a sensory tract (Gowers) which is continued into the cerebrum. This tract is nearly uniform throughout the spinal cord, and is continued upon the side of the medulla to join the fillet. Fourth—internal to the three tracts just described and lying next to the concave outer surface of the grey column of the cord, is a reticulated column composed of the interlacing fibres of the other fasciculi of the lateral column of the spinal cord. This tract is the formatio reticularis of the cord, and corresponds to the corona radiata of the cerebrum and cerebellum. It is also called the mixed lateral column, and is not properly a distinct fasciculus of the spinal cord.

THE MEDULLA OBLONGATA.—The lower portion of the medulla corresponds to the spinal cord in size, shape, and in the arrangement of its columns, with the exception, that the crossed pyramidal tracts of the lateral columns of the cord, pass forward, decussate with each other, and join the direct pyramidal tracts of the anterior columns, to form the anterior pyramids of the medulla. The posterior median columns of the cord are called, in this situation, the posterior pyramids of the medulla, and terminate above in the clavate nucleii at the apex of the fourth ventricle. The posterior column of the cord (Burdach's) occupies the same relative position in the medulla as in the spinal cord, and terminates above in the restiform nucleus, which is situated in the restiform body above and external to the clavate nucleus. The direct cerebellar tract is continued upward, on the surface of

Anterior root zone and sensory tract of Gowers.

Mixed lateral column or the formatio reticularis of the cord.

Anterior pyramids of the medulla. Figs. 9-11.

Posterior pyramids. Figs. 8 10-15.

Posterior or Burdach's column and restiform nucleus. Figs 8 10-26.

the medulla and the restiform body, to the cerebellum. This tract covers the restiform nucleus so that the latter is not exposed upon the surface. The anterior root zone, or lateral tract of the medulla, passes upward upon the side of the medulla to the lower extremity of the olive where it becomes compressed, between the latter and the restiform body, into a narrow tract which above forms the middle layer of the fillet of the pons. The fundamental root zone above the cord is crowded outward by the anterior pyramid, as far as the lower extremity of the olive. Just below this point, it is flattened in front to accommodate the decussation of the pyramids and disappears from the surface of the medulla to get behind the pyramid between the olive and the raphe, in which situation this tract is prismatic in shape, its sides being in relation to the parts just mentioned. Above the olive, it escapes to form a flat layer of fibres upon the fillet, and terminates above in the testes and brachium posterius.

The tracts on either side of the medulla oblongata which form its lower part from before backward are: the anterior pyramid, the fundamental root zone, the anterior root zone or lateral tract, the direct cerebellar tract, the posterior column, and the posterior pyramid.

The upper part of the medulla is attached to the pons Varolii, and is connected with the cerebellum by the restiform bodies, or inferior peduncles of the cerebellum. The medulla gradually enlarges as it ascends from the cord; its upper extremity is about three-quarters of an inch from side to side, and about five-eighths of an inch from before backward. The lower portion of the fourth ventricle is behind the upper half of the medulla oblongata, and the raphe extending from the median fissure of the ventricle forward to the anterior median fissure of the medulla, divides it into lateral halves. A transverse section of the medulla, seen in figure 36, or better in a section one-eighth of an inch above it, is somewhat triangular in shape, with truncated angles, and shows the relation of the tracts which are continued upwards from its lower portion. The relations of these tracts are on either side, beginning from the anterior median fissure and raphe in front: the anterior **pyramid; the** olivary body, or olive; a triangular tract between **the** olive and **raphe, beneath** the pyramid, **called** the fundamental **root zone;** the lateral tract, or anterior root zone, compressed **between the olive and** restiform body; and the restiform **body or inferior peduncle** of the **cerebellum within which is seen the restiform nucleus. Figure 36.**

Relation of the parts of the upper portion of the medulla. Figs. 7 to 15-26.

The parts of the medulla which have not **been described are the** anterior pyramids, the olive, and the restiform body. The arciform fibres and the **fibres from the raphe will be** described incidentally **with** the others above mentioned.

The anterior pyramids **of the medulla emerge from the** middle **of the lower border of the** pons **Varolii as two** rounded cords, **one on each side of the** anterior median fissure and **raphe of the medulla oblongata. They descend** side by side, **separated by the anterior median** fissure, to the point of decussation at the **lower extremity of the medulla** where each divides into two portions: **first, a small tract which continues** on the same side of **the anterior median fissure of the spinal cord, and is** the anterior **median column, or column of Türck; and, a large tract which breaks up into** several bundles of fibres **that pass backward and outward,** interrupting the anterior horns **of the** grey **columns of the cord, and reaching** the anterior surface of the posterior **horns, forms the** crossed pyramidal tract of the opposite side of the **spinal cord.**

Anterior pyramids and division into two parts. Figs. 7 9-11 to 14.

Column of Türck Figs. 11 to 13.

Crossed pyramidal tract. Figs. 14-27.

Arciform fibres.
Figs 7 9.

Near the lower extremity of the olive, the pyramid gives off from its outer side a small tract of fibres, which arch beneath the olive and across the surface of the medulla to the restiform body. These fibres join the restiform body, pass to the cerebellum, and are called the arciform fibres of the medulla. They are probably associated fibres, and do not belong to the pyramidal system which is motor in function. Immediately beneath the pons, fibres are given off from the inner side of the pyramid which enter the raphe to decussate with those from the opposite pyramid, and are continuous on the floor of the fourth ventricle with the roots of the seventh nerve, or its nucleus of the opposite side. Below the olive other fibres enter the raphe to reach the opposite twelfth nerve, or the hypoglossal nucleus.

Roots of the seventh nerve.
Figs. 8 9-10.

Roots of the hypoglossal nerve are between the olive and pyramid.

The anterior pyramid is bounded above externally by the olive, from which it is separated, beneath the pons, by a sulcus, and lower down by a slight groove. The lower portion of the pyramid is in relation externally with the fundamental root zone. This zone above is buried beneath the pyramid, between the olive and raphe.

Olivary body.
Figs. 9 to 14.

The olive is an oblong body situated on either side of the medulla, between the pyramid and the lateral tract, or anterior root zone. It is immediately beneath the pons, from which it is separated by a slight groove. This body is about half an inch in length from above downward, and one-fourth of an inch across its middle. Its upper extremity is connected with a large tract of white fibres which go to form the anterior layer of the fillet proper, and is called the olivary fasciculus of the fillet. The olivary fasciculus is related in front with the fasciculus of the fundamental root zone, and behind with that derived from the lateral tract of the medulla, or anterior root zone of the cord. The olive is covered

Olivary fasciculus.
Fig. 12.

externally by a thin **layer of** transverse striæ, **derived** from the raphe, which passes to the restiform body and to join the commissure of the flocculus. These are probably cerebellar fibres from the nuclei in the floor of the fourth ventricle, belonging to the same system as the uppermost fibres **of the pons Varolii.**

A transverse section **of the** olive (figure 36) shows that it is composed of **an** external **white capsule,** within which is **a** convoluted layer of grey matter surrounding a mass of **white subtance.** From its hilus, which is above and internal to the mass, fibres emerge which form a commissure **between the** opposite olivary bodies, and **others are directed backward to the floor of the fourth** ventricle to **connect** with **the nuclei of the** pneumogastric and hypoglossal nerves. **The upper part** of the olive **with the** former, **and the lower part with the** latter (Van der Kolk). **This body is** supposed to be concerned with **the co-ordination of the muscles of** speech.

The restiform **bodies, or inferior peduncles, connect the** cerebellum with the **medulla and spinal cord. They arise** from the posterior and **outer angles of** the medulla upon the sides of the fourth ventricle, **and arch** upward and **backward between the superior and middle peduncles of the cerebellum, to spread over the upper surface of the corpora dentata, in which they terminate. The upper border of the restiform body forms the posterior side of the restiform triangle,** and is in relation to the parts transmitted by the triangle, viz : the **anterior root of the eighth nerve, the seventh and fifth nerves, and the posterior layer of the fillet. Its posterior border** is in **relation with the posterior root of the eighth nerve and the** commissure of **the flocculus. A** transverse section through the restiform nucleus (figure 36) shows a body

Striæ upon the surface of the olive.

Section of the olive, its hilus, commissure, an l connections with the tenth and twelfth nerves. Fig. 36.

Restiform bodies. Figs. 8 to 18.

Restiform triangle and parts transmitted by it. Figs. 9 to 11.

Commissure of flocculus. Fig. 28.

similar but larger than the nucleus of the olive, and that commissural fibres connect the opposite nuclei.

Restiform system of fibres. Figs 9 10.

The restiform system of fibres of either side is derived: from the restiform nucleus at the side of the lower extremity of the fourth ventricle, from the clavate nucleus, the direct cerebellar tract of the cord, the arciform fibres of the anterior pyramid, the cerebellar fibres from the floor of the fourth ventricle derived from the raphe, and from the uppermost fibres of the pons Varolii, which proceed from the raphe and floor of the fourth ventricle.

RECAPITULATION OF THE TRACTS OF THE CEREBRO SPINAL AXIS.

In the foregoing pages we have studied the several parts of the cerebro-spinal axis in segments, and have connected each by its immediate relations, and by the continuity of the tracts of which it is composed.

It will now be of advantage to review the axis as a whole, and to trace its tracts in continuity through its entire extent, in connection with the nerve centers with which they are associated. In so doing, we will attempt to arrange the several tracts and nerve centers into general systems, to accord with their supposed functions, in order to assist the memory in retaining a knowledge of the anatomy, and to simplify what at first sight appears difficult to the student.

Beginning, therefore, with the spinal cord, we will follow each system from its origin below to its destination above.

The ventricular system begins as a minute canal, extending through the length of the spinal cord and terminating above in the fourth ventricle, at the middle of the posterior surface of the medulla oblongata. The fourth ventricle is a broad cavity, situated behind the upper half of the medulla and the pons Varolii. At its upper extremity the fourth ventricle is contracted into a small canal, the iter, or aquaduct of Sylvius. The iter is about three-fourths of an inch in length, lies beneath the upper extremity of the valve of Vieussens, the commissure of the corpora quadrigemina, and the posterior commissure of the third ventricle. The iter terminates above

The ventricular system.

Central canal of the spinal cord.

Fourth ventricle.

The iter.

in the third ventricle. The third ventricle is a vertical cavity between the thalami optici, extending from the base of the brain to the fornix above, and laterally, upon each side, to the edges of the fornix where it communicates with the lateral ventricles through the fissure of Bichat and the foramina of Monro. The lateral ventricles are large cavities in the hemispheres of the cerebrum, and each ventricle sends projections into the frontal, temporal, and occipital lobes, called the anterior, middle and posterior cornua of the lateral ventricles. The fifth ventricle is enclosed by the two layers of the septum lucidum, and has no communication with the other cavities of the brain. The floor of the third ventricle is connected with a cavity in the pituitary body by a canal through the infundibulum.

The ventricular system communicates with the subarachnoid spaces along the great transverse fissure, and by a foramen in the pia mater above the calamus scriptorius, called the foramen of Majendie.

THE GANGLIONIC SYSTEM: The grey matter of the spinal cord is disposed in two columns, one on each side, enclosed by the white matter of the cord. These columns are united by a grey commissure, which contains in its middle the central canal of the cord, and they are projected forward and backward into cornua, which are the origins of the roots of the spinal nerves. The anterior horns of grey matter are motor in function, and the posterior horns contain the centers of the sensory roots of the spinal nerves. The sensory roots are associated with the sympathetic nervous system by ganglia, near their insertion into the spinal cord. The vesicular columns of Clarke extend along the junction of the posterior cornua and the posterior surface of the grey commissure, one on each side of the posterior median fissure of the cord. The anterior horns of the grey matter are interrupted at the upper extremity of the spinal cord by the decussation of the crossed pyramidal tracts, and above the decussation they gradually approach the median line of the medulla, while the posterior horns gradually diverge or are turned outward.

In the fourth ventricle the grey columns are exposed upon its floor to form the fasciculi teretes; the motor nerve centers being disposed internally along the sides of the median fissure, and the sensory centers upon the outer sides of the ventricle. The grey matter above the fourth ventricle encloses the iter, and contains the centers that control the movements of the eyes. Above the iter, it spreads upon the sides and the floor of the third ventricle; and the sides of the ventricle are united in the center by the grey or middle commissure of the third ventricle. On the floor of the ventricle the grey matter is continuous with the tuber cinereum, infundibulum, and pituitary body, and in front with the lamina cinerea and grey matter of the hemispheres.

The greater ganglia of the cerebro-spinal axis are: the hemispherical, and the external and internal basal ganglia, of the cerebrum; the hemispherical ganglia, the corpora dentata and the ganglia of the pons Varolii, or the internal and external basal ganglia, of the cerebellum.

The lesser ganglia are: the olivary bodies, restiform and clavate nuclei, of the medulla oblongata; the locus niger, corpora quadrigemina, pineal body, the geniculate bodies, the corpora albicantia, and the red nuclei, in and about the crura cerebri.

Other cerebral ganglia are: the claustrum, and the substantia perforata. The olfactory bulbs, though situated in the cranial cavity, may properly be classed with the spinal ganglia.

Fasciculi teretes.

Grey matter of the iter and of the third ventricle.

Greater ganglia.

Lesser ganglia.

Claustrum, substantia perforata.

Olfactory bulb

White substance of the spinal cord.

Of the white substance of the spinal cord, the fasciculi of which it is composed are rearranged as they pass upward through the several portions of the central nervous system, and are associated with the grey masses above mentioned. The direct pyramidal tract, and the crossed pyramidal tract of the opposite side of the spinal cord, unite to form the anterior pyramid of the medulla oblongata. The anterior pyramid enters the pons Varolii and is associated with the reticulated layer of the anterior portion of that body. This layer of the pons has been described as being connected above with the crusta of the crura cerebri, and laterally with the deep layer of the middle peduncles of the cerebellum. It also contains a large amount of vesicular matter, and is the ganglion of the pons Varolii. The crusta is the continuation of the pyramidal tract above the pons, much enlarged by an accession of fibres from this body, and is distributed to the lenticular nucleus, and through the internal capsule, to that part of the hemispherical ganglion situated about the fissure of Rolando, which is the motor area of the hemisphere. The fibres derived from the crusta occupy the middle portion of the internal capsule. The entire tract with its associations, just described, constitute the motor system of the cerebro-spinal axis.

Motor system.

Visual reflex system.

The fundamental root zone lies upon the outer side of the direct pyramidal tract in the cord, and in the medulla as far as the lower extremity of the olive, where it passes inwards beneath the the anterior pyramid between the olive and the raphe. Above the olive, it spreads out into a thin layer of fibres which crosses the fillet obliquely in an outward direction to the upper border of the middle peduncle of the cerebellum, where it becomes superficial upon the anterior part of the outer surface of the superior peduncle, or processus, and terminates in the testes and brachium posterius. These tracts, one on each side, might with propriety be called the posterior peduncles of the corpora quadrigemina, and it is probable that through them, movements of the body, and the maintenance of the equilibrium, are directed by vision. Possibly it is the function of the nates to direct the movements of the eyes to accommodate the vision to the attitudes of the body and the movements of an object, and of the testes to regulate the actions of the muscles to avoid a blow or to grasp an object.

The fundamental root zones, the corpora quadrigemina, brachia, corpora geniculata, the grey matter around the iter and the optic tracts together constitute a visual reflex system.

The anterior root zone, on either side, is divided from the fundamental root zone by the anterior roots of the spinal nerves. It joins the fillet of the pons, in which it lies between the olivary fasciculus in front, and the fillet fibres derived from the floor of the fourth ventricle, behind. In the crus cerebri, the fillet is internal to the crusta, from which it is separated by the substantia nigra. In the crus, it is related internally below with the processus of the same side previous to its decussation, and higher up, with the processus of the opposite side subsequent to its decussation, and, with the red nucleus of the tegmentum. In the internal capsule the fillet crosses the internal surface of the motor tract in an oblique direction, beneath the thalamus, and in front, lies in a broad groove on the anterior margin of the internal capsular fibres of the crusta. It terminates with the fillet fibres derived from the thalamus in the frontal lobe of the cerebrum.

Fillet system.

As the fillet passes upward from the pons Varolii it receives accessions from various ganglia, and a great mass from the thalamus, which unite with it to form the anterior division of the internal capsule. The longitudinal commissures of the cerebrum unite with the fillet in the anterior lobes of the brain, and with it constitute the fillet system. There is evidence to support the theory, that the fillet is the perceptive system through which the mind is made acquainted with its environments; and, through the medium of the longtudinal commissures, with the memories of things and the events of the past, which have been laid up in the temporal and occipital lobes of the brain.

Cerebello-spinal system. The direct cerebellar tract of the spinal cord originates in the vesicular column of Clarke, occupies the outer side of the cord, medulla and restiform body, and with the latter passes into the cerebellum. It probably assists in the formation of a cerebello-spinal system, which is concerned in the function of nutrition.

The posterior median columns are continued in the medulla by the posterior pyramids, and terminate in the clavate nuclei, beyond which their connections are uncertain. If these columns are respiratory tracts, as supposed by some authorities, they probably form an arc with the fifth and pneumogastric nerves, through which the muscles of respiration are co-ordinated and brought into action. Until corrected, we will class this tract and its associations as a respiratory system.

Reflex respiratory system.

The posterior root zone, or Burdach's column, lies external to the posterior median column, and behind the posterior horn of the grey matter of the cord and the lower half of the medulla. This column terminates above in the restiform nucleus, and is continued by the restiform body to the dentate nucleus of the cerebellum. It is associated by the corpus dentatum with the cerebellum behind, and above, by the pro-

Organic and memory system.

cessus with the opposite red nucleus, thalamus, and hemisphere of the cerebrum. The association formed by this tract in conjunction with the cerebellum, and the temporal and occipital lobes of the cerebrum, including the intermediate tracts and grey masses, constitutes the organic and memory system.

Thus the columns of the spinal cord are associated into six systems, having distinct functions, which are co-ordinated at various points by vesicular masses that unite the several systems into one great function, comprehended in the word mind.

The reader is here cautioned against accepting, without due reflection and investigation, the theories advanced in relation to the function of the tracts described and associated into systems. This has been done for the purpose of assisting the memory in retaining the anatomical relations.

CENTRAL ORIGIN AND RELATION OF THE CRANIAL NERVES.

First nerve - Olfactory. Figs 2 7.

The cranial nerves are twelve in number and are enumerated from before backward, as: first, the olfactory, which consists of the olfactory bulb, and nerve or commissure, which is lodged in the olfactory fissure upon the orbital surface of the frontal lobe of either hemisphere of the brain, parallel with the anterior median fissure of the cerebrum. The nerve divides behind into three roots, external, middle, and internal. The external root crosses the outer part of the anterior perforated space to the temporal lobe of the hemisphere, the lower portion of which is supposed to be the cerebral center of taste and smell. The middle root dips into the anterior perforated space, and apparently passes backward to the anterior commissure of the third ventricle. In the mole, an animal in which the sense of smell is highly developed, this root is large and appears to decussate with its fellow in the median line, and passes with the commissure to the temporal lobe of the opposite hemisphere. The internal root passes inward behind the anterior extremity of the marginal convolution and is lost in the median fissure, probably forming, with the internal root of the opposite nerve, the commissure between the olfactory bulbs. The connection of the anterior commissure of the third ventricle with the opposite temporal lobes renders it probable that its function is a commissure between the cerebral centers of smell of the opposite hemi-

CENTRAL ORIGIN AND RELATION OF THE CRANIAL NERVES. 71

spheres. The central arrangement of the olfactory nerves, if the above is correct, is similar to that of the optic nerves.

The second, or optic nerves of each side, unite to form the optic commissure, and again divide into the optic tracts. Each tract winds upon the external surface of the cerebral peduncle, near the posterior part of which it divides into two tracts which terminate: the inferior, in the internal geniculate body; and the superior, a larger tract, in the external geniculate body and pulvinar of the thalamus. The divisions of the optic tract are connected with the corpora quadrigimina by the brachia, anterius and posterius. The posterior fibres of the optic commissure are commissural between the cerebral nerve centers, the middle fibres decussate, the anterior fibres are commissural between the retinae, and the external fibres of each nerve are continued into the tract of the same side.

Second nerve— Optic. Figs. 2-7- 9-10 to 14, 14*.

The third nerve, motor occuli, arises from the inner side of the crus cerebri near the bottom of the posterior perforated space. It pierces the internal margin of the crusta, and upon entering the tegmentum, its fibres spread into numerous striae as they pass through the anterior extremity of the processus, to their destination in the grey matter beneath the iter.

Third nerve— Motor occuli. Figs. 7-9-27.

The fourth nerve, pathetic, arises from the inner margin of the processus, behind the testes, and is connected with the grey matter surrounding the iter which lies immediately beneath its apparent origin.

Fourth nerve— Pathetic. Figs. 8-9.

The fifth nerve, the trigeminus, arises from the center of the side of the pons Varolii. It transfixes the middle peduncle of the cerebellum as it passes obliquely backward, outward, and downward to the inferior angle of the restiform triangle, through which it reaches the grey matter beneath the floor of the fourth ventricle. It divides into several slips which are distributed to several grey nuclei as far down as

Fifth nerve— Trigeminus. Figs. 7 9-10-13.

the calamus scriptorius. Its motor root is small, and at its origin is above and behind its sensory root.

Sixth nerve—Abducent. Figs. 7 9

The sixth nerve, the abducent, lies upon the pons, and, winding beneath the lower margin of this body, passes outward above the olive to join the fillet at its outer margin, where it is in relation with the fifth, seventh, and anterior root of the eighth nerves. Above, it leaves the fillet to reach the grey matter in the floor of the iter.

Seventh nerve—Facial. Figs. 7-8 9 10.

The seventh nerve, facial, arises below the middle peduncle of the cerebellum, internal to the flocculus and the auditory nerve, and, passing upward behind the middle peduncle, between the restiform body and the outer margin of the fillet, it enters the fourth ventricle through the anterior angle of the restiform triangle. It crosses the floor of the ventricle obliquely downward and inward to the raphe, where it decussates with the opposite nerve and joins the anterior pyramid of the opposite side beneath the lower margin of the pons. In the floor of the fourth ventricle it is connected with the nucleus of the seventh nerve. It forms the superior striæ transverse of the floor of the fourth ventricle.

Eighth nerve—Auditory. Figs. 7 to 10.

The eighth nerve, auditory, lies beneath the lower margin of the middle peduncle of the cerebellum, between the seventh nerve and the flocculus. It divides behind into two roots, an anterior and posterior, which embrace the restiform body. The anterior root, in conjunction with the seventh nerve, passes through the restiform triangle, occupying its superior angle, to reach the floor of the fourth ventricle, where it joins the posterior tract of the fillet, and passes downward on the inner side of the restiform body, to reach the auditory nucleus. The posterior root passes inward beneath the arch of the restiform body, and across it to the floor of the fourth ventricle, upon which it forms the inferior striæ transversæ. This root divides into striæ, which are lost in the raphe, and is connected with the arciform fibres, and the commissure of the flocculus. It is probable that the roots of the auditory nerve have the same disposition as those of the olfactory and the optic nerves, and possibly this arrangement is common to the roots of all the sensory nerves.

The ninth nerve, glosso-pharingeal, lies beneath the flocculus above the pneumogastric nerve, and passes inward behind the olive to the floor of the fourth ventricle, where it is connected with its nucleus.

Ninth nerve— Glosso-pharingeal Fig. 7.

The tenth nerve, or pneumogastric, is beneath the ninth, passes inward behind the olive to the floor of the fourth ventricle. Its nucleus is connected with the upper part of the olive by a commissure.

Tenth nerve— Pneumogastric. Fig. 7.

The eleventh nerve, spinal accessory, arises by slips from the lateral tract of the spinal cord as low as the sixth or seventh cervical vertibræ, and from the side of the medulla, which unite into a trunk that lies behind the flocculus and the pneumogastric nerve. Its deep origin is from the anterior horns of the grey matter of the cord.

Eleventh nerve— Spinal accessory. Fig. 7.

The twelfth nerve, hypoglossal, arises from between the olive and anterior pyramid of the medulla. Its roots pass backward between the olive and the fundamental root zone to the floor of the fourth ventricle. Its nucleus is connected with the lower part of the olive and with the opposite anterior pyramid.

Twelfth nerve— Hypoglossal. Fig. 7.

Figure 1. The appearance of the upper surface of the cerebrum, and the convolutions and fissures marked upon it. The great longitudinal fissure separates the hemispheres. The convolutions are named, and the fissure of Rolando indicated upon the left hemisphere; and the three great functions of the brain are located upon the right. The superior extremities of the calloso-marginal and pareto-occipital fissures are marked at their terminations as they emerge from the longitudinal fissure; the former immediately behind the upper extremity of the fissure of Rolando, corresponding to a point just behind the vertex of the skull; and the latter to a point beneath the superior angle of the occipital bone, and the crown of the head. By reference to figure 26, it will be seen that the parieto-occipital and calloso-marginal fissures form the boundaries of certain lobes upon the inner surface of the hemisphere, the upper borders of these lobes occupy the spaces indicated in this plate upon the margin of the longitudinal fissure. The cuneus is located upon surface of the head, by a space extending along the median line from the occiput to the crown; the quadratus—from the crown almost to the vertex; the paracentral lobule—from the vertex to a point an inch and a half in front of it, and the marginal convolution completes the distance to the nasion.

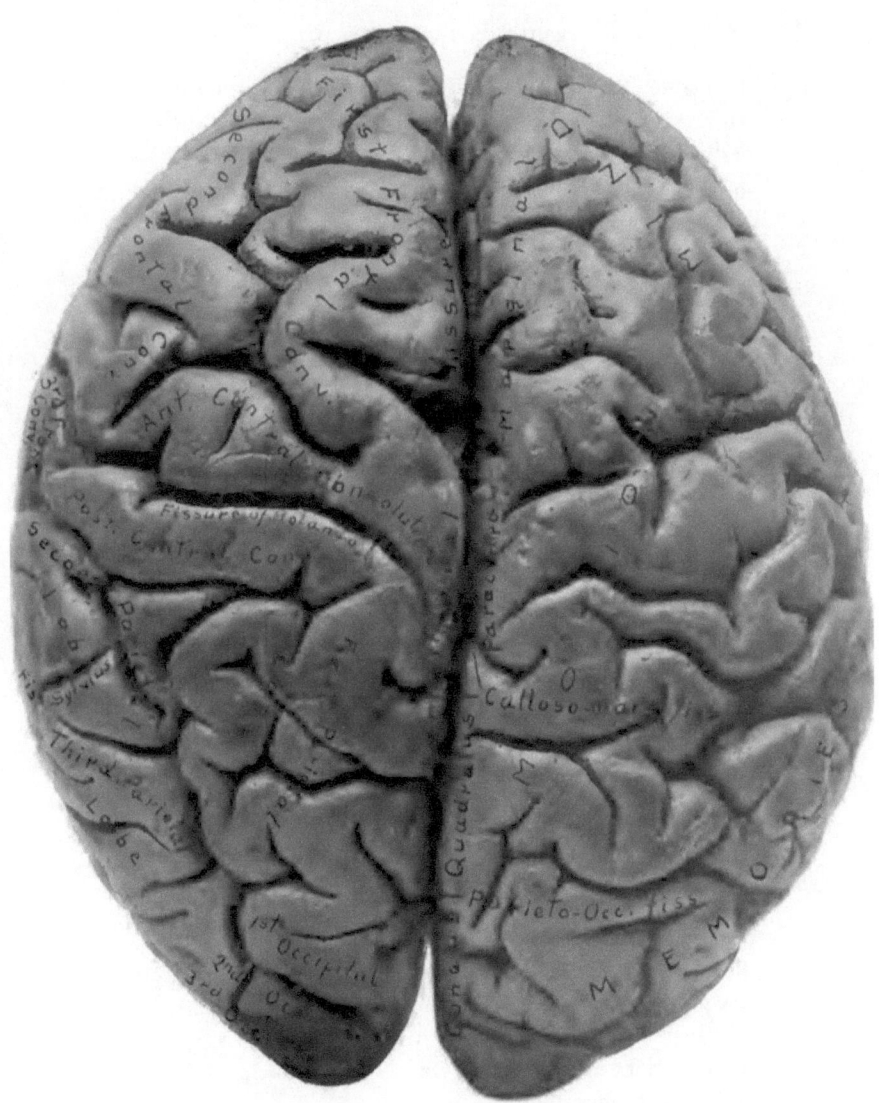

Figure 1.

Figure 2. Base of the cerebrum:—divided into two surfaces, orbital and temporo-occipital. The whole surface is divided into lateral halves from before backward by (1) the anterior median fissure, (2) the interpeduncular space, (3) the oval hilus of the cerebrum, (4) the posterior median fissure. Each lateral half has three lobes, the frontal, and the temporal, separated by the fissure of Sylvius, and behind, the occipital lobe. On each side of the anterior median fissure are the olfactory fissures, bulbs and nerves. Each nerve terminates posteriorly in three roots, external, middle and internal. The space between the temporal lobes contains from before backward: (1) lamina cinerea, obscured by the optic commissure, (2) optic nerves commissure and tracts, (3) the anterior perforated spaces, one on each side of the optic commissure, and leading outward from these spaces are the fissures of Sylvius, (4) enclosed by the optic tracts and the crura cerebri are: the tuber cinereum, infundibulum and pituitary body, corpora albicantia, posterior perforated space, and third nerves. In the hilus of the cerebrum are seen: (1) in the middle line the third ventricle, its anterior, middle and posterior commissures,—the anterior and middle commissures are not visible in the figure,—(2) the pineal body, (3) the splenium and lyra of the corpus callosum. On each side of the middle line is the ruptured end of a crus cerebri or peduncle. The outer side of the broken surface of the peduncle shows: (1) the crusta, (2) in front, the fillet, (3) midway, the bed of the red nucleus, behind which are (4) the ruptured fibres of the brachia of the corpora quadrigemina, and (5) the posterior extremity of the thalamus opticus or pulvinar.

On the margins of the posterior median fissure are (1) the posterior extremity of the gyrus, (2) parieto-occipital fissure, and (3) the lobus lingualis.

The convolutions seen on this surface can be read upon the plate.

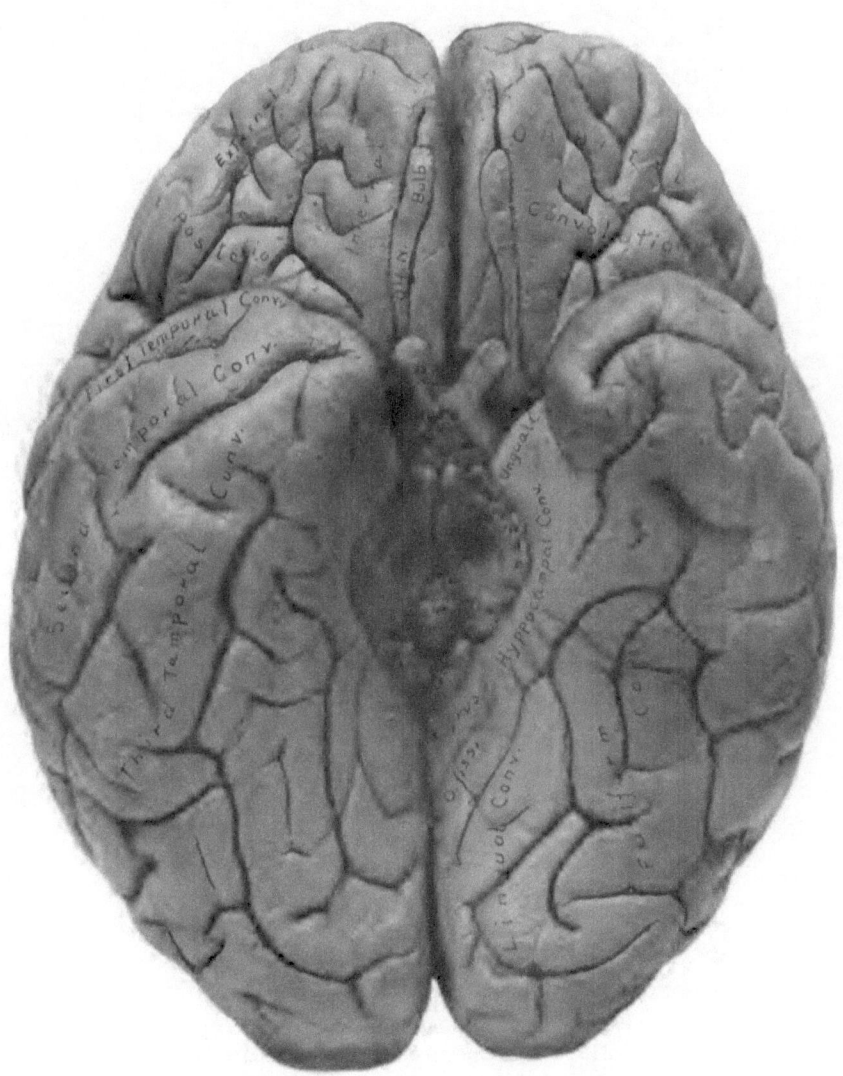

Figure 2.

Figure 3. The upper portion of the cerebrum is removed to a level with the corpus callosum. This body forms a wide furrow in the middle of the surface exposed, and is concave from side to side. Upon each side of the corpus callosum is a crest extending into the anterior and posterior lobes of each hemisphere, which is the line of decussation of the corpus callosum with the internal capsule. Internal to the crest, on each side, is the superior longitudinal commissure, or commissure of the gyrus, connecting the occipital and frontal lobes, represented on the left side of the figure. In the center of the corpus callosum is the raphe, upon each side of which are two longitudinal white lines, the striæ longitudinales or nerves of Lancisci. The corpus callosum is surrounded by the corona radiata, or white substance, in which are seen numerous punctate spots or torn vessels, the puncta vasculosa. The corona is bounded externally by grey matter, the hemispherical ganglia, which is divided in front and behind by the anterior and posterior median fissures, at the bottom of which are lateral extensions, called the ventricles of the corpus callosum. Behind the posterior margin of the corpus callosum are, the gyrus, and the exposed anterior surface of the cuneus. On the lateral margins of the hemispheres are the posterior extremities of the fissures of Sylvius, and a triangle is drawn upon the right side of the figure to illustrate the internuncial fibres, and the external longitudinal commissure; the former connecting the centers of word and sight memories, and the latter associating these centers with that of speech. In front of the fissure of Sylvius is a scheme representing the tracts for automatic movements of the hand produced by visual and aural impressions; and still more anterior, a tract connecting the center for movements of the hand with the speech center, through which the latter center may be excited by movements of the hand, in the event of a lesion of the direct tracts between the aural and visual memories and the speech center.

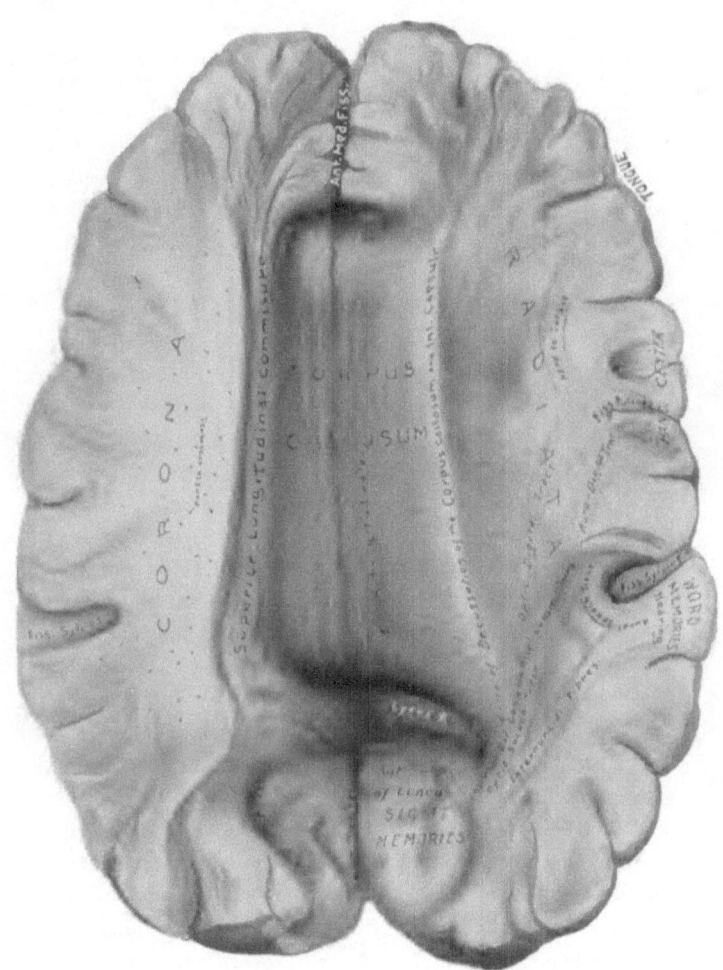

Figure 3.

Figure 4. Fornix and lateral ventricles. In the median line from before backwards are: (1) the anterior median fissure at the posterior extremity of which is a lateral fissure, the ventricle of the corpus callosum, (2) the genu of the corpus callosum, (3) septum lucidum, containing the fifth ventricle, (4) the fornix, the anterior pillars of which bending downward bound the foramena of Monro in front; upon the upper surface, in the median line, is an elevated ridge which forms the attachment of the fornix to the under surface of the corpus callosum, (5) the splenium of the corpus callosum, posterior to which is a part of the ventricle of the corpus callosum, and (6) the posterior median fissure.

On each side of the median line are the lateral ventricles, basal ganglia and internal capsules, corona radiata and the hemispherical ganglia. The lateral ventricle is divided, into a body, an anterior, middle, and posterior cornua. The ventricle is bounded externally in front by the internal capsule, upon which the functions of its parts are indicated. Behind the capsule is a broad groove leading from the body of the ventricle to the middle and posterior cornua. The posterior portion of the capsule bends downward and is called its genu. The body of the ventricle contains from within outward; (1) the fornix and corpus fimbriatum, (2) fissure of Bichat, (3) choroid plexus, (4) part of the thalamus opticus, (5) tænia semicircularis, (6) caudate nucleus. The anterior cornu contains the head of the caudate body; the middle cornu, the hippocampus major, and pes hippocampi; the posterior cornu contains the hippocampus minor; and between these cornua is the pes accessorius or eminentia collateralis. The tail of the caudate body, tænia semicircularis, choroid plexus, and corpus fimbriatum descend together along the inner margin of the middle cornu to the apex of the ungual convolution. See Figs 5, 7, 17. The hemispherical ganglia constitute the grey matter which encloses the corona radiata.

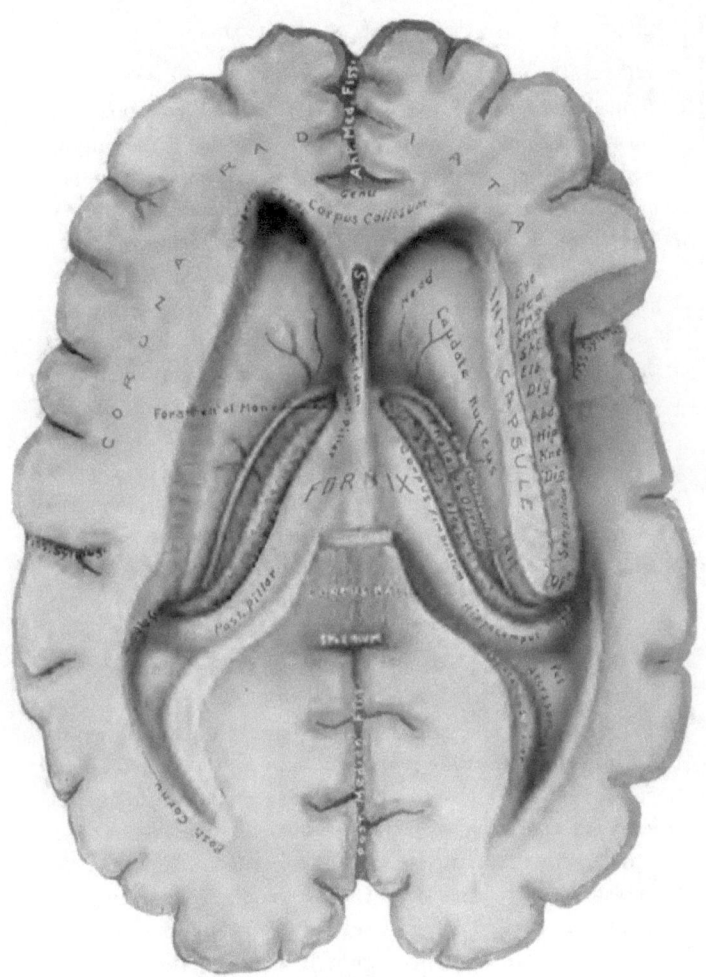

Figure 4.

Figure 5. A horizontal and lateral view, with the fornix divided and reflected, exposing the velum interpositum. In the median line from before backward are seen: (1) the anterior median fissure, (2) genu of the corpus callosum, (3) septum lucidum and fifth ventricle, (4) anterior pillars of the fornix, (5) velum interpositum, triangular in shape, fringed by the choroid plexuses, and containing the venæ Galeni and choroid arteries, and, at its center behind, an elevation denoting the position of the pineal gland. (6) Behind the velum is the reflected fornix and corpus callosum, upon which are seen transverse striæ, the lyra. The attachment of the fornix to the corpus callosum is seen in the middle of the section of these parts. Behind the corpus callosum, is the posterior median fissure. The parts forming the floor of the lateral ventricle, as well as the internal capsule, are shown in figure 4. The oblique view presented by this figure exposes fully the middle and posterior cornua and the parts contained in them, also the situation and extent of the corpus fimbriatum, transverse fissure, and choroid plexus, as well as the manner in which they encircle behind the genu of the internal capsule into the middle cornu. A complete view of the pes hippocampi and pes accessorius is exposed. At the lower and anterior extremity of the middle cornu is a transverse section of the lenticular nucleus, external capsule, claustrum, insula and Sylvian fissure. Below the lenticular body, or external basal ganglion, is a tract of transverse fibres, which is a section of the inferior ramus of the internal capsule. See figure 7. The inferior ramus of the internal capsule forms the roof of the middle cornu. In the roof of the middle cornu which has been removed in this dissection are the tænia semicircularis and the tail of the caudate nucleus lying parallel to the optic tract. See figure 7.

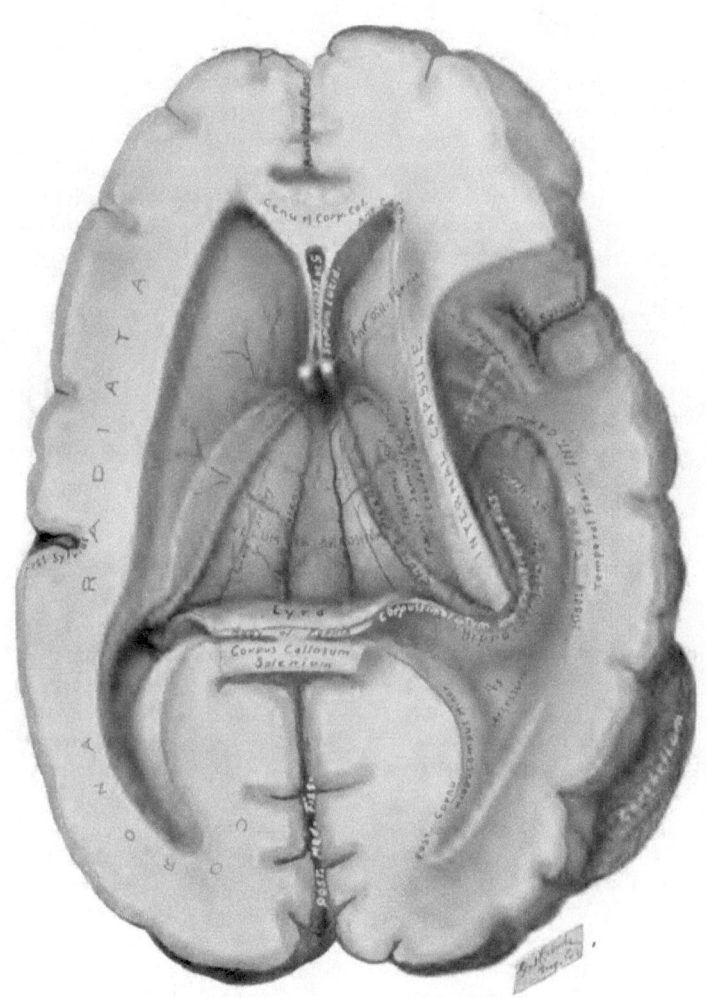

Figure 5.

Figure 6. In this dissection, the velum interpositum and the right occipital and temporal lobes are removed, exposing the third ventricle, pineal gland, corpora quadrigemina, basal ganglia, and the upper surface of the cerebellum. Figure 7 is a lateral view of the same dissection, and shows the relation of the external basal ganglion, or lenticular nucleus, to the internal capsule; also the posterior extremities of the caudate nucleus and tænia semi-circularis, as they descend behind the genu of the internal capsule into the roof of the middle cornu of the lateral ventricle.

In the median line from before backward (Fig. 6) are seen: the anterior median fissure, genu of the corpus callosum, septum lucidum enclosing the fifth ventricle, the divided anterior pillars of the fornix, the third ventricle containing three commissures, anterior, middle, and posterior, the pineal gland and its pillars, corpora quadrigemina divided into four tubercles by a crucial depression, the incisura anterio of the cerebellum, the superior vermiform process, and the incisura posterior. Laterally: in front, is the corona radiata; in the middle, the internal and external basal ganglia, separated by the internal capsule; and behind, the upper surface of a lateral lobe of the cerebellum, made up of the lobus quadratus and the posterior, or semilunaris lobe. The internal basal ganglia are, the caudate nucleus and the thalamus opticus, divided by the tænia semicircularis, which lies in a depression between them. The thalamus is composed of four distinct bodies; (1) the pison (pea), a small body at the junction of the anterior pillar of the pineal gland with the body of the thalamus, (2) the pulvinar, pointed in front opposite the middle commissure, and continuous behind with the external geniculate body and outer division of the optic tract, (3) the tuberculum, a nodule at the anterior exteremity of the thalamus, continuous with a slim tract of white fibres directed backward and outward, along a groove on the upper surface of the thalamus which lodges the corpus fiimbriatum, (4) the fusiformis, which is an oblong elevation lying on the outer border of the thalamus, internal to the tænia semicircularis.

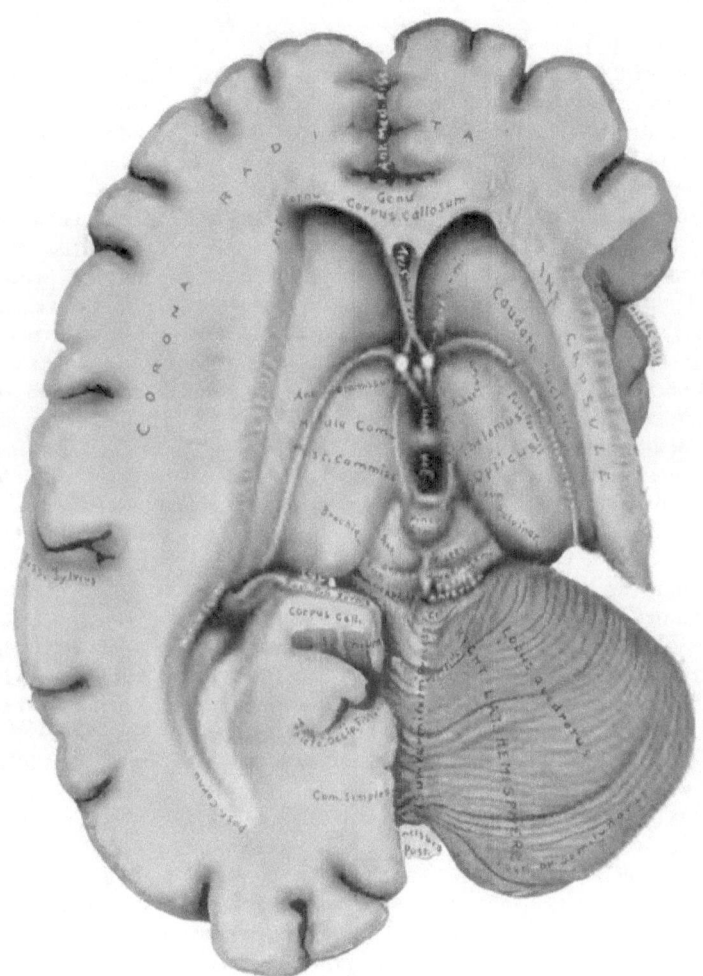

Figure 6.

Figure 7 is a lateral and inferior view of the same dissection as figure 6, from which may be studied on the right of the figure (left side of the brain) the parts seen upon the base of the brain. On the left side of the figure are seen; the tracts of the medulla oblongata, the expansion of the crus cerebri into the internal capsule, and the relation of the capsule to the lenticular nucleus, to the optic tract, tænia semicircularis, and tail of the caudate nucleus. The pons is divided to show the descent of the crusta, and the continuation of a portion of its fibres into the anterior pyramid of the medulla. The under surface of the cerebellum is also exposed, and in conjunction with figure 6 the parts of this portion of the brain can be studied. The following parts are seen on the base of the brain from before backward in the median line: (1) anterior median fissure, (2) lamina cinerea (hidden by the optic commissure), (3) optic commissure, nerves and tracts, (4) interpeduncular space, bounded by the optic tracts and cerebral peduncles, and containing, the tuber cinereum, infundibulum and pituitary body (removed), corpora albicantia, posterior perforated space and third nerves, (5) the pons Varolii, (6) the medulla oblongata. On the outer side from before backward are: (1) the orbital convolutions, (2) olfactory bulb, and nerve terminating in three roots, (3) the anterior perforated space, upon which is the anterior pillar of the corpus callosum, (4) the fissure of Sylvius, (5) temporal lobes, (6) crus cerebri and fourth nerve, (7) great transverse fissure, and (8) the flocculus and under surface of the cerebellum. The cranial nerves are numbered from before backward: 1st, olfactory; 2nd, optic; 3rd, motor oculi; 4th, pathetic; 5th, trigeminus; 6th, abducent; 7th, facial; 8th, auditory; 9th, glosso-pharyngeal; 10th, pneumogastric; 11th, spinal accessary; 12th, hypoglossal.

On the left side of the figure behind, is the incisura posterior and inferior vermiform process, lobes of the cerebellum, tracts of the medulla, arciform fibres, olive, crusta, optic tract and geniculate bodies, and the inferior extremities of the tænia semicircularis, tail of caudate body ending in a bulbar extremity, and the auricular expansion of the fibres of the internal capsule enclosing the lenticular nucleus. Upon the surface of the lenticular body is a fragment of the external capsule, at the inner angle of which is (AC) the outer extremity of the anterior commissure of the third ventricle.

The internal capsule is divided into four parts: the anterior, which is hidden by the frontal lobe; the middle, which is the motor portion; the posterior, containing the fibres of common sensation and the optic radiation; and inferior ramus of the capsule to the temporal convolutions, which contains the tracts of hearing, taste, and smell.

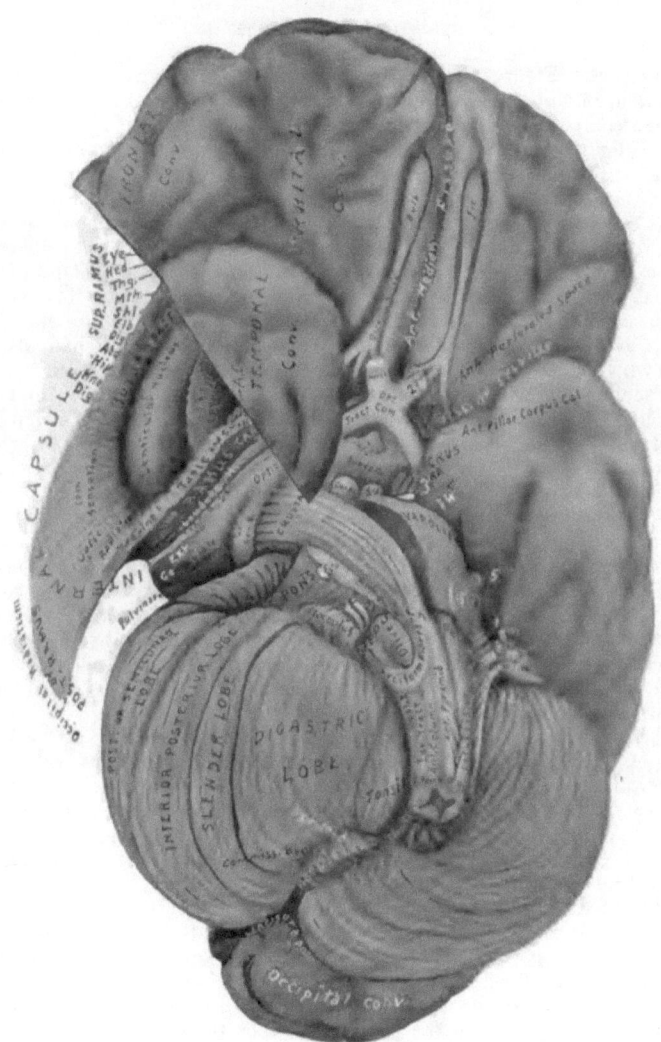

Figure 7.

Figure 8. Figures 8 and 9 are posterior and lateral views of the same dissection, in which the hemispheres and the cerebellum have been removed giving a complete view of the intermedia.

This figure can be divided into three sections: a superior or cerebral, a middle or peduncular, and an inferior or medullary

At the upper, or anterior part of the superior division is a part of the corpus callosum crossing above the anterior horns of the lateral ventricles, and uniting at its lateral extremities with the internal capsules. The corpus callosum and the capsules enclose, as in a triangle, the ventricles and the internal basal ganglia, caudate nuclei and thalami optici. The caudate nucleus of the left side has been removed in order to show the distribution of fibres from the thalamus to the internal capsule, and also the fibres which emerge from between the lamina of the capsule to reinforce the tænia semicircularis as it passes forward into the anterior horn of the lateral ventricle. The fornix lies over the third ventricle, and, in its middle is joined to the under surface of the corpus callosum by the septum lucidum, which contains the fifth ventricle, and separates the anterior horns of the lateral ventricles. The posterior part of the thalamus winds around the crus cerebri to be continuous with the optic tract. The opening to the third ventricle is surrounded by the pineal body and its anterior peduncles, and upon each side in the front of the pineal body is a small tubercle of grey matter, called the pison. At the posterior extremity of the thalamus are two eminences, one on each division of the optic tract, the corpora geniculata, externum and internum. Between the thalami are four tubercles, divided by a crucial depression, the corpora quadrigmina, or nates and testes. Directed obliquely outward from the nates and testes to the geniculate bodies are the brachia, anterius and posterius.

The middle division contains from within outward: the valve of Vieussens, processes and fourth nerves, fundamental root zones and crustæ. A vertical groove upon each crus cerebri lies between the root zone and the crusta. Lower down, on either side, the relations of the parts are from the median line outward: the valve of Vieussens, processus, corpus dentatum, restiform body, and the pons Varolii. These latter parts form an arch or roof over the fourth ventricle.

The inferior division is broad above, and flattened beneath the arch formed by the parts just mentioned, where it presents a triangular depressed surface which is part of the floor of the fourth ventricle. This surface is crossed by the roots of the seventh and eighth nerves, the striæ transversæ. In the median line is the posterior median fissure of the fourth ventricle, which terminates below in the central canal of the spinal cord, and is continuous with the posterior median fissure of the medulla oblongata. Bounding the sides of the floor of the fourth ventricle are the restiform bodies, and at the apex of the ventricle are two eminences on each side, the anterior are the restiform nuclei, and the posterior, the clavate nuclei. The lower portion of this division is rounded, the columns of which it is constituted are similar to those of the spinal cord, a section of which is seen on the end.

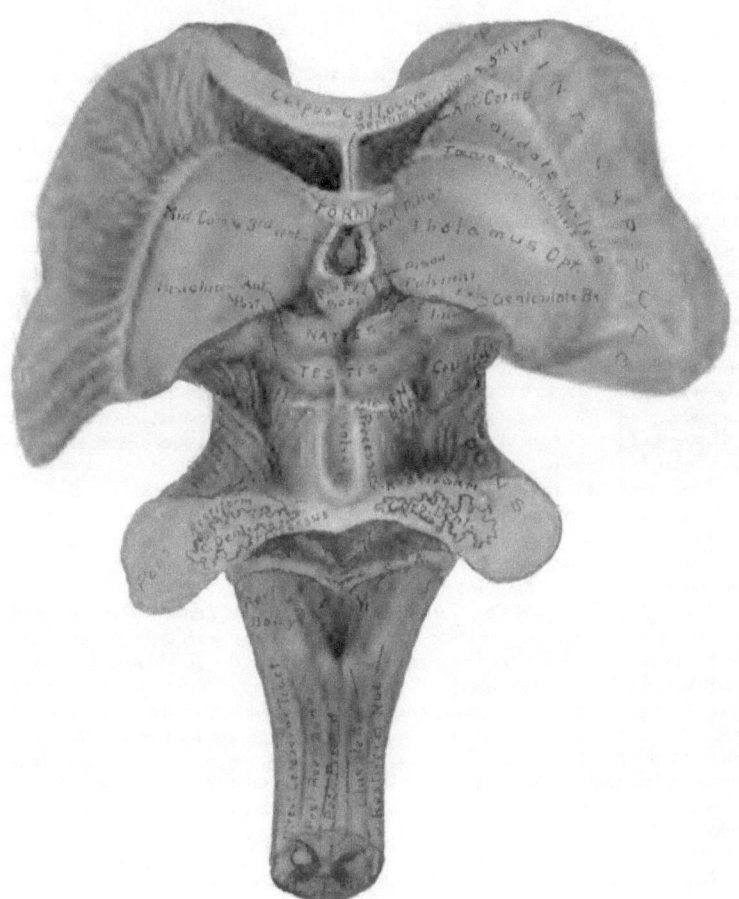

Figure 8.

Figure 9. For the convenience of description, figure 9 may be divided into four portions: cerebral, peduncular, pons, and medullary.

In the center of the cerebral portion is the lenticular nucleus, or external basal ganglion, which is bounded above by the deep fibres of the external capsule, the superficial fibres having been removed in order to expose the ganglion. The external capsule is derived from the outer surface of the ganglion, and its fibres radiate upwards in the middle, and in front and behind, they are directed obliquely forward and backward (See Fig. 10). The internal capsule receives the fibres of the external capsule, radiates forward where it is called the fillet, upward in the middle, which are its motor fibres, and behind and below, which are the sensory and memory divisions of the capsule. The external longitudinal commissure extends from the sensory radiation of the internal capsule to its fillet division, which it joins in front of the lenticular nucleus. The commissure lies beneath the lenticular body and upon, or rather within the inferior, or temporal radiation of the internal capsule. Beneath the longitudinal commissure and the inferior ramus of the internal capsule are, from behind forward: the corpora quadrigemina, geniculate bodies, the upper and lower divisions of the optic tract, the optic tract commissure and nerves, and the lamina cinerea.

The peduncular division contains from behind forward, the processus, the fourth nerve, fundamental root zone, vertical groove, crusta cerebri, and interpeduncular space containing the tuber cinereum, infundibulum, corpora albicantia, posterior perforated space and third nerves. The latter parts, together with the optic commissure and lamina cinerea form the floor of the third ventricle.

The region of the pons Varolii contains from before backwards: (1) the superficial or commissural layer, the upper fibres of which are derived from the raphe and floor of the fourth ventricle, and enter the cerebellum along with the restiform body, to be distributed to the corpus dentatum; (2) the reticulated layer, or formatio reticularis of the pons, composed of the fibres of the crusta, anterior pyramid of the medulla, and the deep layer of the pons, interspersed with grey matter, and is the nucleus of the pons, or external basal ganglion of the cerebellum; (3) the fillet, composed of four layers: fundamental root zone, olivary fasciculus, anterior root zone, fillet tract from fourth ventricle; (4) the fifth, seventh and eighth nerves, which pass through the restiform triangle (see figures 10-12) into the floor of the fourth ventricle, the roots of the latter two nerves form the striæ transversæ on the floor of the ventricle (figures 8-10); the sixth nerve winds beneath the pons and joins the fillet; (5) the reflected superior portion of the arch of the restiform body, corpus dentatum and posterior root of the auditory nerve.

The medulla oblongata in its upper half is composed of parts from the median fissure backward: the anterior pyramid overlying the fundamental root zone, the olivary body, anterior root zone, and the restiform body, composed of the arciform fibres of the anterior pyramid, the direct cerebellar tract, and the posterior root zone. The restiform body contains the restiform ganglion at its lower extremity. Burdach's column, the posterior pyramid and clavate nucleus, are behind, and are not seen in the figure (see figure 8). The lower part of the medulla contains the same parts except the olive. The decussation of the anterior pyramids are in front, about midway from the pons and the lower end of the figure. The lower extremity of the decussation marks the limit of the medulla oblongata, below which is the spinal cord, the columns of which can be read upon the figure.

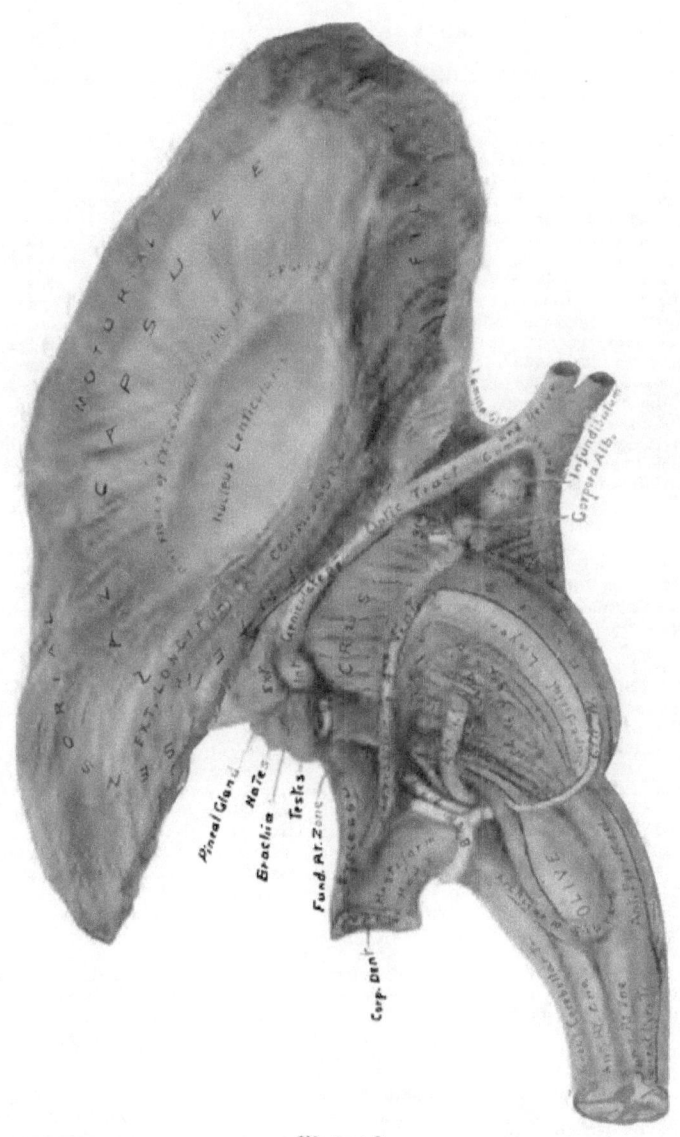

Figure 9.

Figure 10. Dissection of the pons and basal region of the cerebrum, viewed from behind. The parts upon the right half of the figure are retained in order to show the relations of the parts uncovered by the dissection upon the left side. In the median line in front is a section of the genu of the corpus callosum and septum lucidum, the anterior pillars of the fornix, the third ventricle, the fasciculi teretes and posterior median fissure of the iter and fourth ventricle, the calamus scriptorius, clavate nuclei, and the posterior median fissure of the medulla oblongata and spinal cord. The calamus scriptorius (pen) is situated between the restiform bodies, and its point is between the clavate nuclei, at the conjunction of the posterior median fissure of the medulla, and that of the floor of the fourth ventricle with the upper extremity of the central canal of the spinal cord (compare with Fig. 26, at R).

Upon the right side of the figure, the parts seen from before backward are: a portion of the corpus callosum and the internal capsule, the anterior horn of the lateral ventricle, caudate nucleus, tænia semicircularis, the thalamus opticus and anterior pillar of the pineal gland, the pison, posterior commissure, pineal body, corpora quadrigemina and brachia, the processus, valve of Vieussens, and corpus dentatum. Beneath are: the fourth ventricle at its widest part, the arch of the restiform body, the restiform triangle, through which the roots of the seventh nerve and the anterior root of the eigth nerve emerge into the fourth ventricle, the former crossing the floor of the ventricle to enter the raphe, and the latter to join a tract, internal to the restiform body, which passes from the auditory nucleus of the medulla to the fillet. The tract just mentioned forms the posterior layer of the fillet (see figure 13, designation 4), and divides the restiform triangle into two portions, an anterior part, which transmits the roots of the seventh and eighth nerves, and an inferior part, through which the fifth nerve passes in front of the fasciculi teretes. Beneath the arch of the restiform body is the posterior root of the eighth nerve, which crosses the floor of the fourth ventricle to reach the raphe. This root of the eighth and that of the seventh are the striæ transversæ on the floor of the ventricle. At the lower extremity of the restiform body is the restiform nucleus (see section of restiform nucleus, figure 36). The clavate nucleus is internal and below the restiform nucleus and is superficial, while the latter is imbedded beneath a layer of white fibres. The posterior pyramid of the medulla, Gall's column, terminates above in the clavate nucleus, and the posterior root zone, column of Burdach, in the restiform nucleus.

On the left side, from before backward are: portions of the caudate nucleus and thalamus, anterior fibres which pass forward to the internal capsule in conjunction with the fillet of the pons Varolii, and are distributed to the anterior lobe of the cerebrum. The red nucleus, or subthalmic ganglion is exposed, and is seen in relation with the fasciculi teretes and the wall of the third ventricle internally, and with the fillet externally. The red nucleus gives off fibres to the thalamus and in front to the fillet, and behind, a large cord continuous with the opposite processus, which decussates with the processus of the same side beneath the fasciculi teretes, under the floor of the iter. External to the red nucleus, the processus and its decussation, are successively from within outward: the fillet, locus niger, and the crusta above; and the fundamental root zone and pons Varolii below. Below the crusta are the uppermost fibres of the pons Varolii, which form a tract derived from the floor of the fourth ventricle, that entering the raphe, passes forward to the posterior perforated space, and winding around the front of the crus cerebri, is directed backward to the restiform body, with which it enters the cerebellum. Below the posterior extremity of this tract, is seen the divided extremity of the restiform body enclosed by the roots of the eighth nerve, the seventh nerve passes through the restiform triangle in company with the anterior root of the eighth, to enter the fourth ventricle. Above the restiform body, the parts from within outward are: the fasciculi teretes and floor of the fourth ventricle, the four layers of the fillet, and the deep and superficial layers of the pons Varolii. The word fillet is written upon the fundamental root zone (see figure 11). The fifth nerve is seen passing under the floor of the ventricle between the deep layer of the fillet and the anterior root zone. The parts forming the posterior and lateral surfaces of the medulla can be read upon the figure.

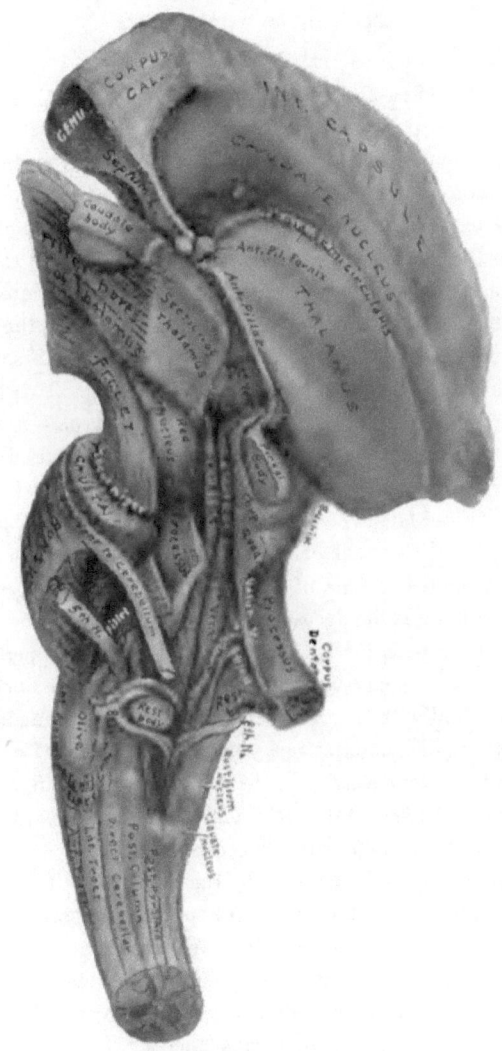

Figure 10.

Figure 11—Is a dissection in which the right pyramid of the medulla, the right half of the pons, and the lower portion of the crusta are removed, exposing the continuity of the fundamental root zone, from the spinal cord to its termination in the testes and brachium posterius. In the cord, it is seen to lie external to the direct pyramidal tract (column of Türck); in the medulla as far above as the olive, it is superficial and external to the anterior pyramid. Opposite the olive, it is compressed into a prismatic shape (figure 36) between this body and the raphe. Upon the anterior surface of the fillet and processus, it is a flat band of fibres terminating above and behind in the testes and brachium posterius. This tract is probably a part of the visual reflex system, by which a person affected with ataxia is able to maintain the upright position while the eyes are open.

The lower extremity of the figure shows the superficial tracts which form the anterior and lateral columns of the cord, viz.: the column of Türck and the fundamental root zone of the anterior column; and the anterior root zone and direct cerebellar tract, of the lateral column of the cord. In the middle of the medulla is the decussation of the pyramids and a flattening of the fundamental zone for its accommodation. The upper part of the medulla is seen to be connected with the pons and the cerebellum by: (1) the anterior pyramid with the reticulated layer; (2) the fundamental root zone, olivary fasciculus, anterior root zone, and fasciculus from the fourth ventricle forms the lemniscus or fillet; (3) the grey matter of the cord forms the fasciculi teretes (not seen); (4) the restiform body connects the posterior part of the medulla with the cerebellum.

The restiform triangle is enclosed, by the edge of the fillet in front, the restiform body behind, and the processus above. The parts transmitted by the triangle are seen in figures 9 and 10. Above the pons are from before backward: the optic nerves, commissure and tracts, interpeduncular space, crusta, locus niger, geniculate bodies, brachia and corpora quadrigemina. Above the optic commissure is the lamina cinerea.

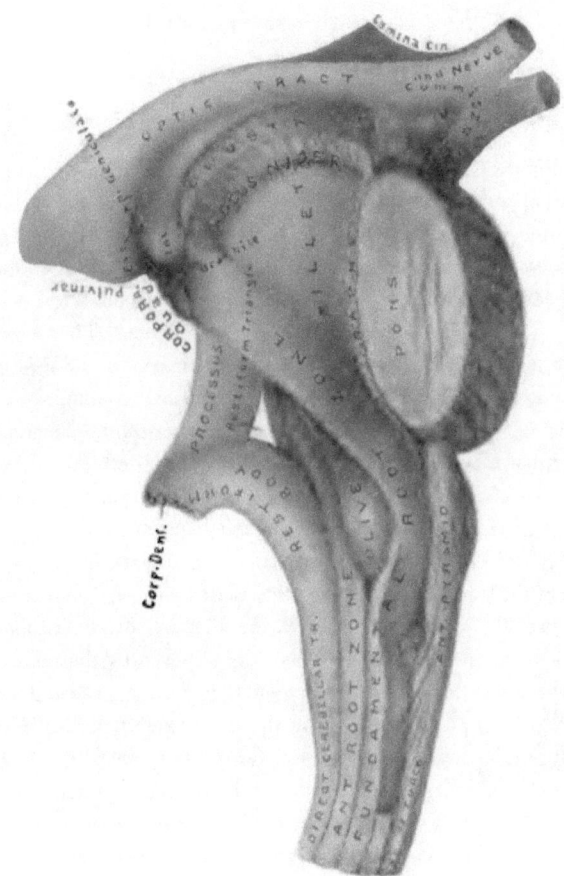

Figure 11.

Figure 12—may conveniently be observed in four divisions. The upper portion, above the optic tract, shows the three divisions of the lenticular nucleus, surrounded by the internal capsule, upon which is marked its divisions, into fillet, motor, and sensory portions. The anterior part is connected with the optic commissure by the lamina cinerea. The second portion of the figure is between the lower arm of the internal capsule and the upper border of the pons; containing the optic nerves, commissure, tracts, and the corpora geniculata, the tuber cinereum and infundibulum, corpora albicantia, posterior perforated space, crusta, locus niger, upper extremity of the fundamental root zone, brachia, corpora quadrigemia, and pineal body. The third portion contains: in front, a section of the transverse fibres and reticulated part of the pons, and a part of the raphe; further back, the fillet proper, consisting of the olivary fasciculus, (which is fully exposed by the removal of the fasciculus of the fundamental root zone), the anterior root zone and a fasciculus from the inner side of the restiform body, derived from the floor of the fourth ventricle. Upon the external margin of the fillet is the restiform triangle, contained behind by the restiform body and the processus, which terminate in the corpus dentatum of the cerebellum. Above and below the fillet, are seen the divided ends of the fundamental root zone. The lower portion of the figure shows the parts which form the medulla oblongata, the decussation of its anterior pyramids, and a section of the spinal cord.

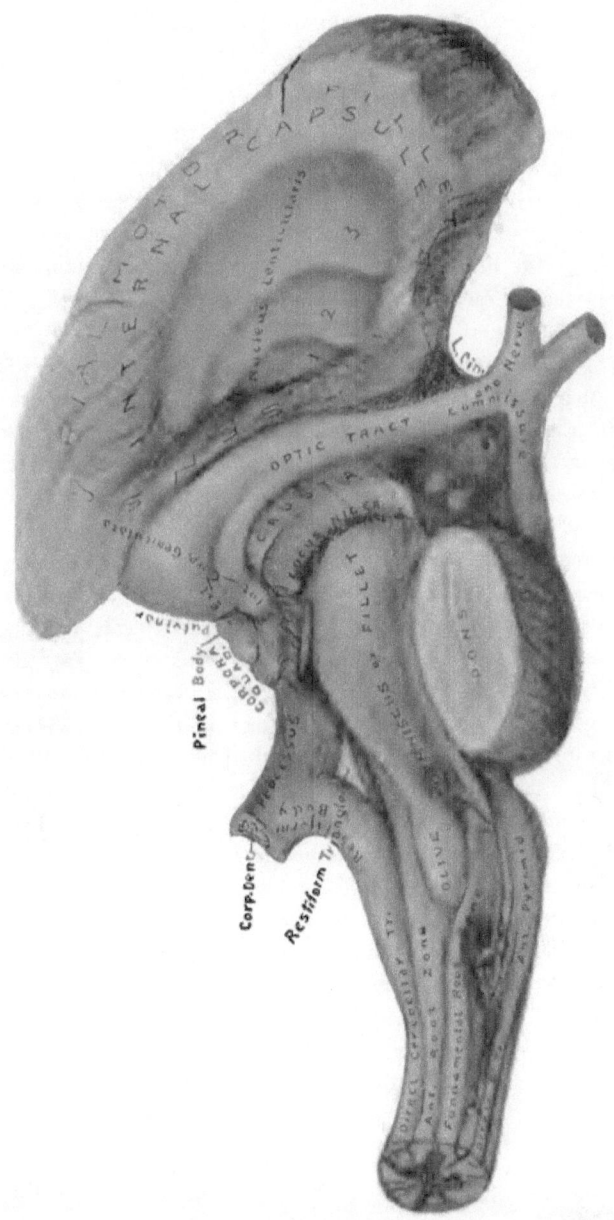

Figure 12.

Figure 13. In this figure, the right pyramid of the medulla, and the right half of the pons, and crusta, have been removed. Sections of the four layers of the fillet have also been removed. The lower ends are marked: (1) the fundamental root zone; (2) olivary fasciculus; (3) the anterior root zone, or lateral tract of the medulla; (4) layer from the floor of the fourth ventricle, beneath which is seen the anterior surface of the fasciculi teretes. Internal to these layers is the raphe, and a section of the pons, consisting of a reticulated and a commissural layer. External to the fillet is the restiform triangle, restiform body, and processus; (5) the fifth nerve. The processus arises behind from the corpus dentatum, is directed upward beneath the fundamental root zone and the fillet, and inward in front of, or under the fasciculi teretes, to decussate with the opposite processus in the raphe of the crus cerebri, (see P T, figure 26). Above the decussation, the fibres of the opposite processus are seen to terminate in the red nucleus, which is uncovered in the figure by the removal of a section of the fillet. Above the red nucleus the fillet is seen to pass upward and forward, beneath the fibres of the internal capsule derived from the crusta, to join the anterior portion of the capsule. The anterior commissure of the third ventricle, in its passage outward to the temporal lobe, pierces the anterior border of the fillet. The three divisions of the internal capsule are indicated upon the figure.

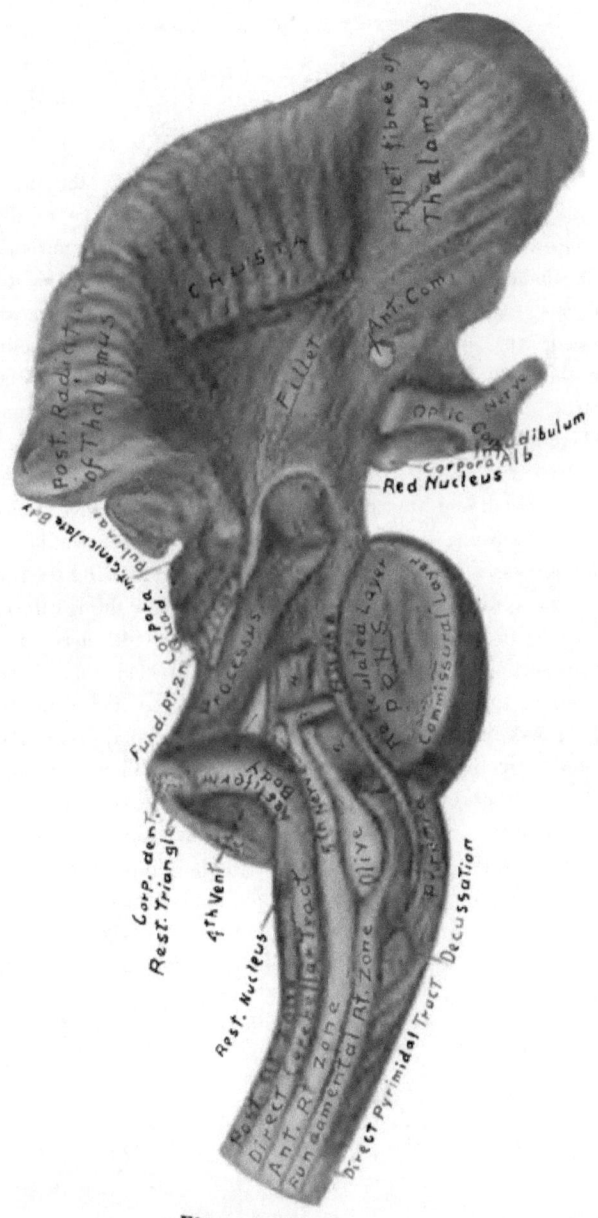

Figure 13.

Figure 14 illustrates from above downward: the inner layer of the internal capsule; the lining membrane of the lateral ventricle which forms the bed of the caudate nucleus; the thalamus opticus, composed of concentric layers of fibrous matter between which vesicular layers are interposed; the divided extremity of the anterior commissure of the third ventricle; the lamina cinerea, optic tract, and the parts in the interpeduncular space; the red nucleus, from which a tract passes forward to the fillet; behind the red nucleus are the corpora quadrigemina, from which the brachia pass outward and upward beneath the geniculate bodies to the internal capsule; below the red nucleus is the decussation of the processes, through the raphe of the crura cerebri; the processus, fillet, and the restiform body enclose the restiform triangle, in front of which are: the root of the fifth nerve, the deep portion of the fasciculi teretes, the divided ends of the layers of the fillet, the raphe and pons; below the pons are the tracts of the upper portion of the medulla, except the anterior pyramid, which has been removed; the lower extremities of these tracts have also been removed, in order to expose the crossing of the fibres of the opposite anterior pyramid to form the deep fibres of the lateral column of the cord. The fibres of the crossed pyramidal tract are seen to interrupt the anterior grey column in their passage from the pyramid to the opposite lateral column of the cord.

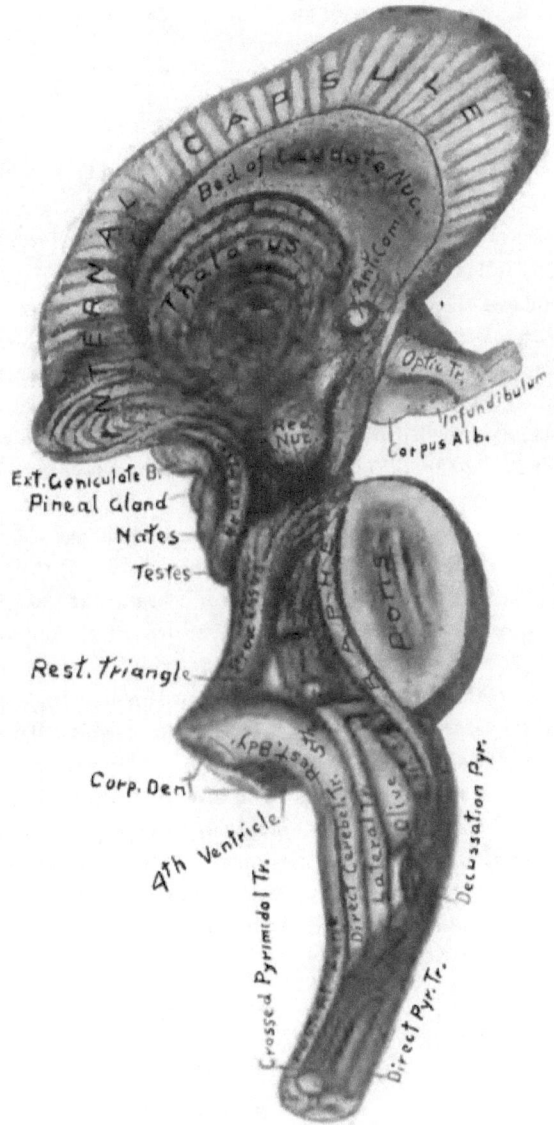

Figure 14.

Figure 15—Beginning from below, shows: (1) a section of the spinal cord, and of the columns of grey and white matter, commissure, and central canal; (2) the columns upon the posterior surface of the medulla, the clavate, and the position of the restiform nuclei; (3) the fourth ventricle, fasciculi teretes, calamus scriptorius, and the parts which bound the fourth ventricle, viz.: the restiform bodies and the processes; (4) the dentate bodies, and the relation of the peduncles of the cerebellum; the superior peduncle enters the under surface of the dentate body, the inferior arches upward and backward between the superior and middle peduncles, and is also connected with the corpus dentatum, while the middle peduncle is external, and has no connection with that body; (5) the processes, above, are seen to decussate beneath the fasciculi teretes, each to connect with the red nucleus of the opposite side; (6) between the red nuclei are the third ventricle, and middle commissure. From within outward on either side are: the third ventricle and the grey matter upon its sides, the red nucleus, fillet, locus niger, and crusta. The fundamental root zone is seen to be in apposition with the posterior margin of the fillet; its divided extremity has been exposed by the removal of the corpora quadrigemina. (7) The grey columns of the cord are spread out upon the floor of the fourth ventricle and continued through the iter to the third ventricle by the fasciculi teretes.

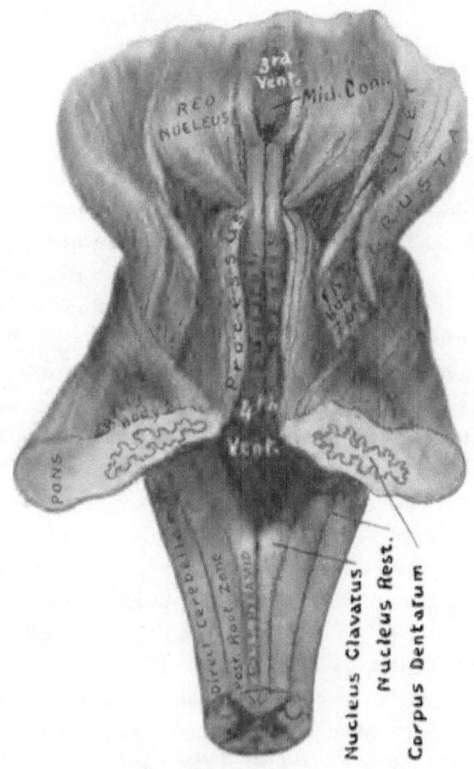

Figure 15.

Figure 15½ represents a dissection in which the upper portion of the cerebellum has been removed by separating the sides of its transverse fissure and dividing the superficial layer of the middle peduncle. The upper portion is torn off from behind forwards, exposing the corona radiata and the upper surface of the corpus dentatum. The deep fibres of the middle peduncle are seen to pass directly into the corona radiata, to be distributed to the hemispherical ganglion of the cerebellum. The superior peduncle passes beneath the corpus dentatum to enter its hilus; and the inferior peduncle, or restiform body, springing up from between the superior and middle peduncles, arches backward and is distributed to the upper surface of the dentate body. The corona radiata is made up of radiating fibres from the middle peduncle, and from the corpus dentatum. The corpora quadrigemina, and fasciculi teretes of the iter, have been removed in order to show the decussation of the processes and their connection with the red nuclei. Above the decussation is the raphe, on either side of which is the red nucleus, fillet, locus niger, and crusta. Behind the crusta is the vertical line of the crus cerebri and the divided extremity of the fundamental root zone. Below the decussation are the divided ends of the fasciculi teretes and the iter, the valve of Vieussens, and, lower down, between the corpora dentata, is the white substance of the middle lobe of the cerebellum.

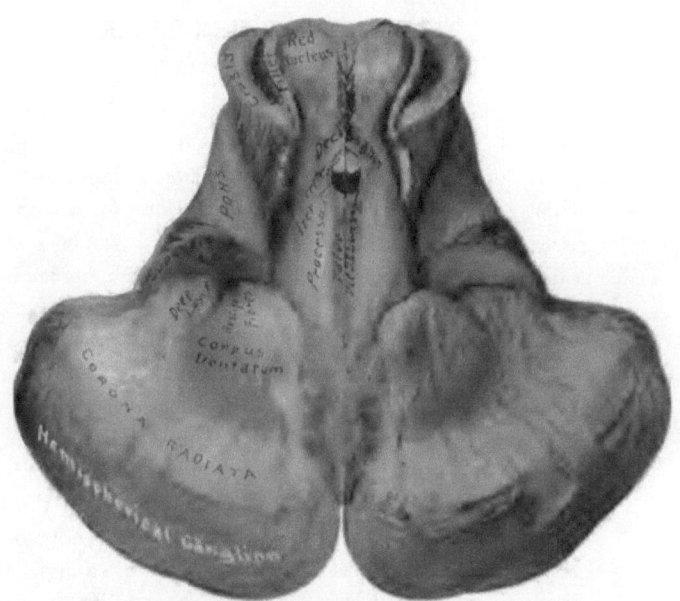

Figure 15½.

THE CEREBRAL HEMISPHERE.

Separated from the remaining portion of the axis, the cerebral hemisphere in general view resembles the quarter section of an ovoid. It is elongated from before backwards, bluntly pointed at each extremity, and presents for examination, three borders which enclose three surfaces. The borders are: superior, external, and inferior. The superior border forms the margin of the great longitudinal fissure, its convexity is upward, and it extends from the posterior extremity of the occipital lobe to the anterior angle of the orbital surface of the frontal lobe of the hemisphere, dividing the external from the internal surface. The external border arches around the anterior border of the orbital surface of the frontal lobe to the fissure of Sylvius, and around the outer border of the temporal and occipital lobes to the posterior extremity of the hemisphere, where it joins the superior border at an angle. This border is interrupted by a broad notch opposite the fissure of Sylvius, which receives the posterior arched border of the orbital plate of the frontal and sphenoid bones. Behind the projection of the temporal lobe this border is arched upward over the cerebellum. It forms a notch in front with the cerebellum which lodges the petrous portion of the temporal bone, and marks the position of the external auditory meatus. This border separates the external and the inferior surfaces. The inferior border is internal along the median line of the base of the cerebrum and is in three portions. The anterior and posterior portions form the margins respectively of the anterior and posterior median fissures. The middle portion crosses the hilus of the hemispherical ganglion, dividing it into

Superior border. Figs. 1 15 17-24.

External border. Figs. 2 7-16-17-24 25.

Inferior border. Figs. 2-17 25.

two unequal parts: that below, **assists** in the formation **of the hilus of** the cerebrum, and that above, is called the intrafissural **portion of the hilus, because it** is surrounded by the great **longitudinal fissure.** Crossing this border at the junction of its anterior and **middle portion is a broad groove continuous** laterally with the anterior perforated space, **and in front with** the anterior surface of the corpus callosum. **This groove is** converted into a foramen **by the optic nerve, and lodges the** anterior cerebral artery. **The bottom of the groove is convex from** within outward, resembling a large cord, and is named the cordon, because, in the early development of the brain this part forms the connection of **the hemisphere to the parts** below it, and because, **the convolutions of the cerebrum can be traced** as loops, **which begin and end in this locality.** The cordon **is composed of a large mass of antero-posterior** fibres disposed in **horizontal** laminæ.

Cordon. Fig. 17.

The surfaces of the hemisphere are: external, **inferior, and** internal. The external surface is pointed before and behind, **and convex in every** direction, corresponding to the concavity **of the inner surface** of the **vault of the cranium,** with which it lies in contact. **Behind the center of** this surface is an oval eminence corresponding to **the parietal eminence of the** skull, which is situated vertically above the external auditory meatus. The vertex is **midway** between **the occiput and the nasion, and is in a direct** line **with the parietal eminence and the external auditory meatus. The relations** of these points to the **external surface of** the hemisphere are of great assistance in locating the lobes and fissures beneath the surface of the head.

External surface.
Figs 16-24-25.

This surface **of the hemisphere is marked by two prominent fissures, the fissures** of **Sylvius and of Rolando; and by minor fissures that separate the** convolutions and the lobes, **and are named after the parts** which they divide.

Fissures of Sylvius and Rolando.
Figs. 16-24.

The fissure of Sylvius begins at the anterior perforated space, winds in front of, to the outer side, and above the temporal lobe, where it divides into a short and long arm; the former projecting upward and forward into the frontal lobe, and the latter is directed backward to the middle of the parietal eminence. This fissure is very deep and branched at its bottom to enclose a bunch of convolutions, called the insula, or island of Riel. The insula is exposed by separating the margins of the fissure of Sylvius, and the convolutions which cover it are called the operculum—a name which must not be confounded with the aperculum, or hilus of the hemispherical ganglion. The position of the fissure of Sylvius is indicated upon the surface of the head by a line drawn from the outer canthus of the eye to the middle of the parietal eminence, and if continued this line would pass through the crown and the opposite external auditory meatus.

The fissure of Rolando extends from the middle of the posterior arm of the fissure of Sylvius, upward and slightly backward to the vertex, dividing the external surface of the hemisphere into anterior and posterior parts, from which it derives the name of central fissure. Its situation is indicated upon the surface of the head as lying beneath the upper portion of a line extending from the inferior angle of the malar bone to the vertex. The convolutions upon the external surface of the hemisphere are represented upon figure 16, and are named: first, second and third, in the several regions of the surface, viz.: frontal, parietal, occipital, and temporal, with the exception of the central region, in which there are only two convolutions, the anterior and posterior central, one on each side of the central fissure of Rolando; and the insular region consists of five or six small convolutions continuous with those of the sides of the fissure of Sylvius.

The inferior surface of the hemisphere, before described with the under surface of the cerebrum, requires a short review in this connection. It is divisible in three portions: an anterior or orbital portion, which is triangular, concave from side to side, and elevated upon its internal border, where it is marked by a longitudinal fissure which lodges the olfactory bulb and nerve. It is marked by three indistinct convolutions: anterior, external, and posterior. The temporo-occipital surface looks obliquely inward, is concave from before backward, and is marked by several convolutions, the outermost of which is the under surface of the third temporal convolution, internal to which in succession are: the fusiformis, lingualis, and hippocampal convolutions. The latter is continuous behind the splenium of the corpus callosum, with the convolution of the gyrus, and in front with the ungual, a small convolution which is turned inward upon the margin of the hilus cerebri. The remaining portion of this surface consists of a crescent depression which includes the interpeduncular space and the divided surface of the crus cerebri, around the outer margin of which is the fissure of Bichat.

The internal surface of the hemisphere presents two portions for examination: a convoluted portion above, in front, and behind; and an enclosed portion, presenting the divided surfaces of commissures and the lateral wall of the third ventricle, which together fill up the upper part of the aperculum or hilus of the hemispherical ganglion, before called the intrafissural portion of the hilus. The convoluted portion of the internal surface of the hemisphere encloses the outer surface of the corpus callosum, from which it is separated by the fissure of the gyrus fornicatus, or ventricle of the corpus callosum. It is wider behind than in front, and shaped like the falciform process of the dura mater, against which it lies; its posterior border, which is the margin of the posterior median fissure, rests upon the tentorium cerebelli. This portion of the internal surface is marked by two distinct fissures, the calloso-marginal, and parieto-occipital. The anterior fissure, calloso-marginal, begins in front beneath the genu of the corpus callosum, and, winding in front and above this body is directed upward and backward to terminate upon the margin of the great longitudinal fissure behind the vertex and the upper extremity of the central fissure of Rolando. The calloso-marginal fissure is separated from the corpus callosum by the convolution of the gyrus fornicatus. The parieto-occipital fissure begins below the hippocampal convolution upon the inferior surface of the hemisphere, extends upward and backward, across the internal border and internal surface, to a point upon the margin of the longitudinal fissure which lies beneath the crown of the head. Above the inferior or internal border of the hemisphere it gives off a branch fissure, the calcarine, which passes backward to the posterior extremity of the occipital lobe. These divisions of the parieto-occipital fissure inclose a wedge shaped bunch of convolutions, called the cuneus, which is the center of sight memories. The convolutions seen upon this surface from behind forward are: the inner margin of the lingualis, the cuneus, quadratus, paracentral, marginal and gyrus fornicatus.

The intrafissural portion of the internal surface of the hemisphere is bounded in front by the rounded surface of the cordon, and from below backward, the ungual and hippocampal convolutions which are continued above and in front of the corpus callosum by the gyrus fornicatus, which terminates at the cordon in front. The surface included within these boundries is filled up by sections of the corpus callosum, fornix, septum lucidum, anterior, middle and posterior commissures of the third ventricle, lamina cinerea, optic commissure, tuber cinereum and infudibulum, corpora albicantia,

THE CEREBRAL HEMISPHERE.   111

raphe of the crus cerebri, **the** decussation of the processes, the iter, corpora quadrigemina, **and** pineal body. It contains the outer wall of the third ventricle, formed by the inner surface of the thalamus opticus; and, extending from the splenium, beneath the corpus callosum and fornix, to the anterior pillars of the fornix in front, is the fissure of Bichat, the anterior extremity of which is the foramen of Monro.

If the above parts are removed, and also the crus cerebri, there is seen a large oval opening, leading into the center of the hemisphere, which is the aperculum, or hilus of the hemispherical ganglion. The grey matter of the ganglion terminates at the margin of the aperculum, and is plainly seen at its lower part, where it is called the fascia dentata. Internal to its upper margin is a tract of longitudinal fibres, called the superior longitudinal, or commissure of the gyrus. The cordon is seen to be composed of horizontal layers of longitudinal fibres, and about half an inch above its lower surface is a groove, seen in figure 17, which is the bed of the anterior commissure of the third ventricle. Attached to the apex of the ungual convolution are the lower extremities of the corpus fimbriatum and the tænia semicircularis, between which is seen the lower extremity of the fissure of Bichat leading to the middle cornu of the lateral ventricle.

The convolutions of the hemisphere which compose the hemispherical ganglion **can be recollected in** the following **order:** external, internal, and posterior orbital; insula; first, second, and third frontal; anterior, and posterior central; first, second, and third parietal; first, second, and third occipital; first, second, and **third temporal;** fusiformis; lingualis; ungual, hippocampal, and **gyrus fornicatus; cuneus, quadratus, paracentral** and marginal.

Marginal notes: Aperculum, or hilus of the hemisphere. Fascia dentata. Commissure of gyrus. Cordon. Lower extremities of the corpus fimbriatum, tænia semicircularis and fissure of Bichat. Convolutions comprising the hemispherical ganglion.

Figure 16 is a representation of the external surface of a hemisphere and the convolutions and fissures upon it. The principal fissures are: (1) the fissure of Sylvius, commencing below at the junction of the anterior and middle third of the lower border, and extending upward and backward, terminates in the middle of the parietal eminence, and a short anterior arm is directed upward and forward into the frontal lobe; and (2) the fissure of Rolando, commencing below about the middle of the posterior arm of the fissure of Sylvius, extends upward and backward to the vertex, dividing the upper portion of the of the hemisphere into anterior and posterior halves, hence this fissure is often called the central fissure. The convolutions upon this surface are: orbital, frontal, central, parietal, occipital and temporal. The orbital convolutions are: external, posterior and internal—seen in figure 2. The first frontal convolution forms the outer surface of the marginal convolution (figure 17), joins the orbital below, and the upper extremity of the anterior central above. The second frontal has two portions; the upper portion joins the middle part of the anterior central, and the lower portion joins the posterior part of the third frontal. The third frontal surrounds the anterior arm of the fissure of Sylvius. The anterior central convolution is continuous with the marginal above and the insula below. It will be seen that these convolutions form arches, which through the orbital and and insular convolutions are connected with the cordon (figure 17). The posterior central is connected with the insula below, dna the quadratus above, and behind with the parietal lobes. This system includes the first temporal, parietal convolutions, posterior central, quadratus, gyrus and hippocampal convolutions, completing arches from the cordon in front. Of the remaining convolutions, they are seen to join the lower temporal and hippocampal convolution at angles, and are continued by them to the cordon. It will be observed that this arrangement of the convolutions of the hemisphere is admirably adapted to the passage of longitudinal fibres from every part of the surface of the brain to the frontal lobes, which are the seat of the intellect and judgment.

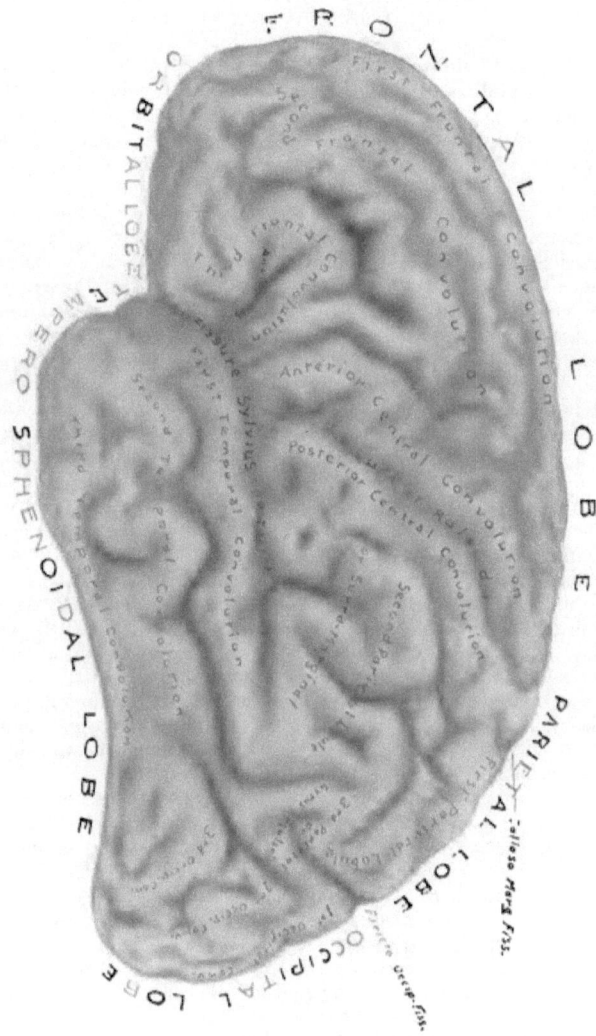

Figure 16.

Figure 17 shows the internal surface of the hemispherical ganglion and the convolutions of which it is composed. Its interior is exposed by the removal of the corpus callosum, fornix, septum lucidum, basal ganglia, and the crus cerebri; parts of which fill up the hilus or aperculum of the ganglion. There are two principal fissures upon this surface: (1) the calloso-marginal, beginning at the cordon below, and, arching above the gyrus fornicatus, terminates upon the margin of the great longitudinal fissure between the paracentral lobule and the quadratus, a little behind the upper extremity of the fissure of Rolando; (2) the parieto-occipital fissure commences upon the lower surface of the hemisphere external to the hippocampal convolution, and passing upward and backward above the internal margin of the hemisphere divides into two arms which enclose the cuneus. The anterior arm terminates in the margin of the longitudinal fissure at a point beneath the superior angle of the occipital bone, the crown; the posterior arm passes backward to the extremity of the occipital lobe, and is called the calcarine fissure. The convolutions and their boundaries can be seen upon the figure, and also the connections by which they are continued to the cordon. The grey matter of the hemisphere terminates within the margin of the aperculum, and, being distinctly seen below is called the fascia dentata. Within the upper margin of the aperculum is a large tract of longitudinal fibres, extending from the cordon to the hippocampal convolution, which is the superior longitudinal, or commissure of the gyrus. The cordon is a large collection of longitudinal fibres, disposed in horizontal layers, underlying the anterior portion of the internal capsule and basal ganglia. Its under surface is rounded and exposed on the base of the brain at the anterior perforated space. It is perforated by large vessels which supply the basal ganglia. The anterior commissure of the third ventricle passes through it in a lateral direction, and its bed, uncovered, is seen in the figure. The ungual convolution is seen turning backward behind the cordon, and, attached to its apex are the tænia semicircularis and corpus fimbriatum, between which is the lower extremity of the fissure of Bichat entering the middle cornu of the lateral ventricle from the under surface of the brain. The parts which fill up the aperculum are seen in figure 26. The upper part of the aperculum is interhemispherical; the lower portion transmits the crus cerebri, and is part of the hilus of the cerebrum.

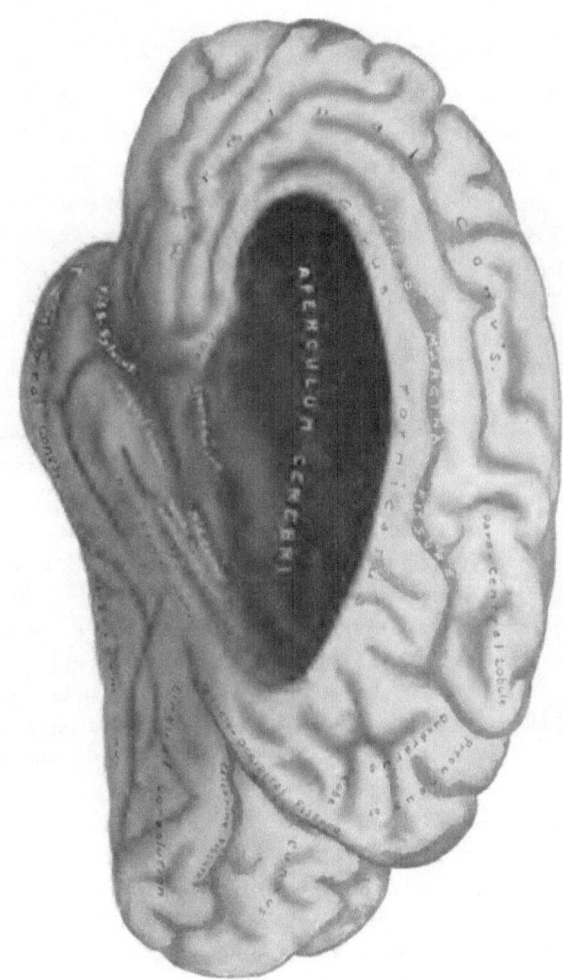

Figure 17.

Figure 18. The insula is exposed by the removal of the convolutions forming the margins of the fissure of Sylvius. The convolutions removed are called its operculum (cover), and must not be confounded with the aperculum or hilus of the hemispherical ganglion (figure 17).

The insula, or island of Riel, is composed of six or seven small convolutions, which are continuous above and below with those of the margins of the fissure of Sylvius. The anterior extremities of the convolutions forming the insula are all directed toward the cordon, which is at the bottom of the fissure of Sylvius and occupies the anterior perforated space (figure 17). Above the insula is an arch, called the vault, composed of superimposed horizontal layers of fibres, derived from the corpus callosum and the internal capsule. This disposition of the fibres, derived from these bodies, permits the passage of longitudinal fibres between them in their passage outward to the convolutions of the surface. The same arrangement of fibres from these bodies is seen below the insula, between which the fibres of the external longitudinal commissure are interposed in their passage from the posterior to the anterior lobes of the hemisphere. Behind the insula are fibres from the thalamus passing backward between the plates or layers of the internal capsule. The internal capsule encloses the insula, which is the outermost part that fills up the funnel shaped concavity enclosed by the capsule. Beneath the convolutions of the hemisphere is the internuncial layer of short fibres, which extend between adjacent convolutions.

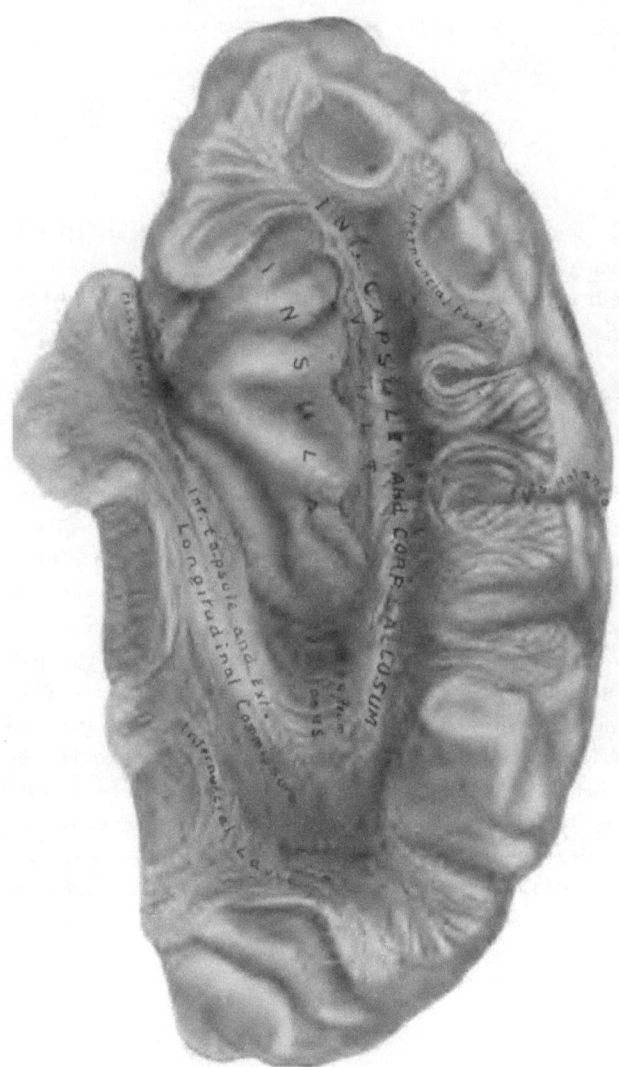

Figure 18.

Figure 19. Upon the removal of the insula and the subconvolutional white layer which subtends its grey substance, a broad and thin layer of grey matter is uncovered, called the claustrum, which is difficult to represent in a lateral dissection, but is to be seen in a vertical section, (see vertical sections from figures 10 to 18). This layer of grey matter is thicker below than above, and covers the external surface of the external capsule, with which it is in immediate relation. This layer having been removed, the dissection here represented shows the external capsule to be composed of fibres radiating upward from beneath the external longitudinal commissure. This layer of fibres is very thin immediately above the commissure and thick beneath the vault. Its fibres are derived from the outer surface of the lenticular nucleus, and pass upward to unite with the under surface of the fibres of the vault, into which they are lost. The external longitudinal commissure is distinctly seen to form a part of the cordon and to enter the anterior part of the internal capsule in front of the lenticular body. Its posterior extremity intermingles with the posterior fibres of the corpus callosum and internal capsule, and with them is distributed to the posterior convolutions of the hemisphere.

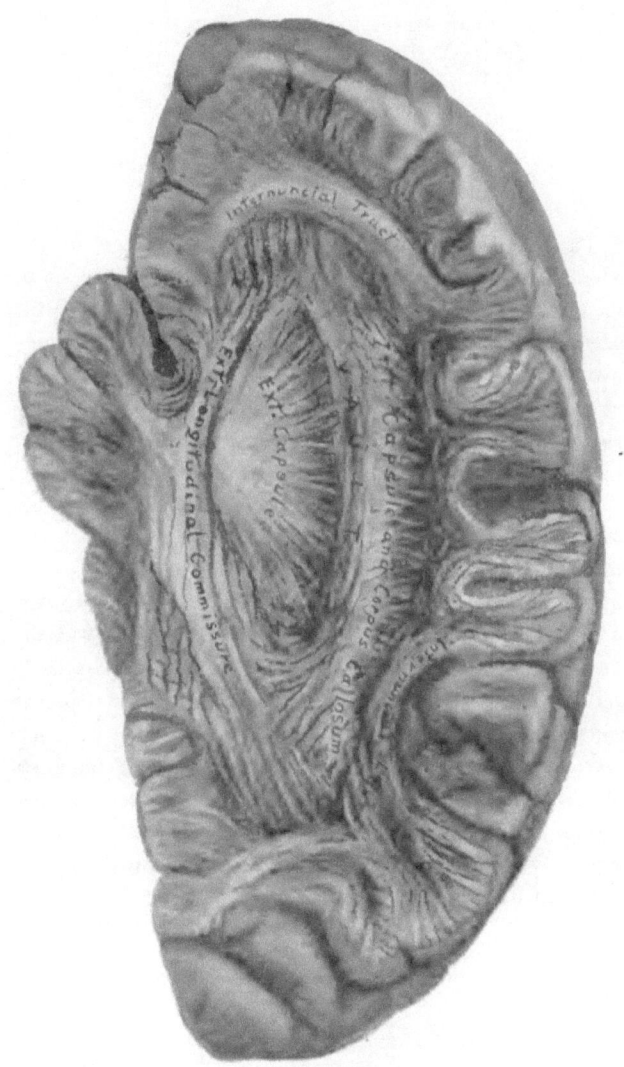

Figure 19.

Figure 20. In this dissection the outer fibres of the external capsule are removed to expose the external division of the lenticular body (putamen). A small fragment of the external capsule is seen above the external longitudinal commissure, and the deeper portion of it is seen above the exposed portion of the lenticular nucleus. These latter fibres pass directly upward from the surface of the nucleus to enter the vault.

In front of the lenticular body, the external longitudinal commissure joins with the anterior portion of the internal capsule; and behind, its fibres interlace with the posterior and inferior radiation of the capsule. Between the posterior part of the lenticular nucleus and the genu of the internal capsule, the posterior fibres from the thalamus emerge to pass backwards to the occipital lobes. These fibres from the thalamus contain the optic radiation to the cuneus and the posterior convolutions of the hemisphere. The radiation of the fibres derived from the internal capsule and corpus callosum, forming the vault, surrounds the lenticular nucleus, except a space opposite to the fissure of Sylvius, which is occupied by the longitudinal commissure and cordon.

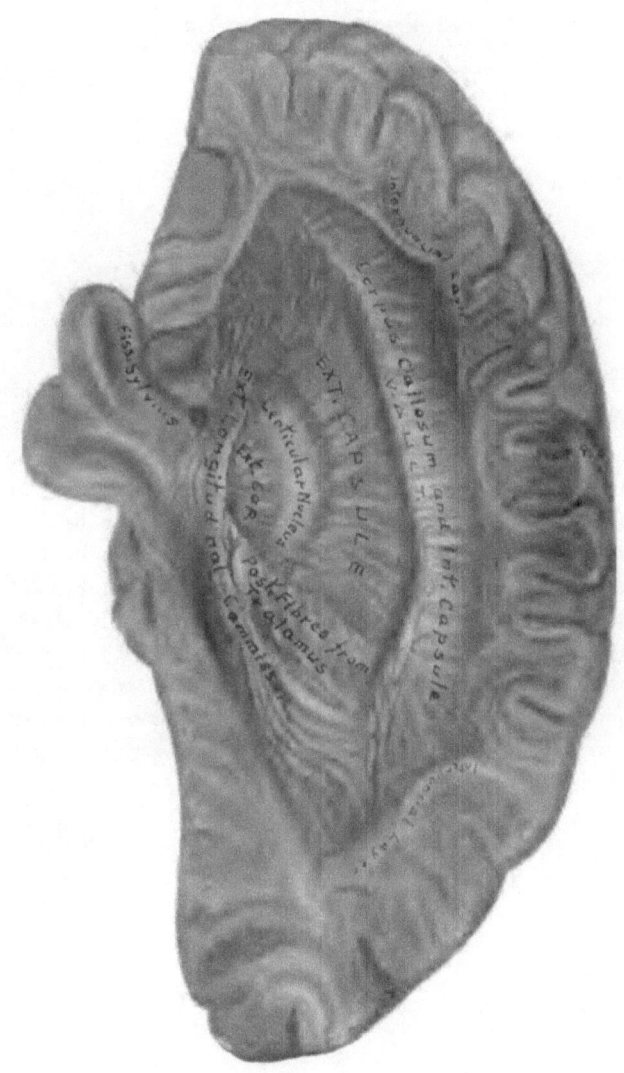

Figure 20.

Figure 21. The lenticular nucleus and upper portion of the crusta fibres of the internal capsule have been removed, exposing the anterior and posterior distribution of fibres from the thalamus. The fillet is seen to emerge from internal to the crusta, and to pass forward to join the anterior distribution of fibres from the thalamus. The external longitudinal commissure is fully exposed, and is seen to unite with the fillet fibres in front after forming the mass of fibres which constitute the cordon. The anterior commissure of the third ventricle passes obliquely outward and backward, across the depression which lodges the lenticular body, to reach the outer extremity of the inferior ramus, or temporal distribution, of the internal capsule.

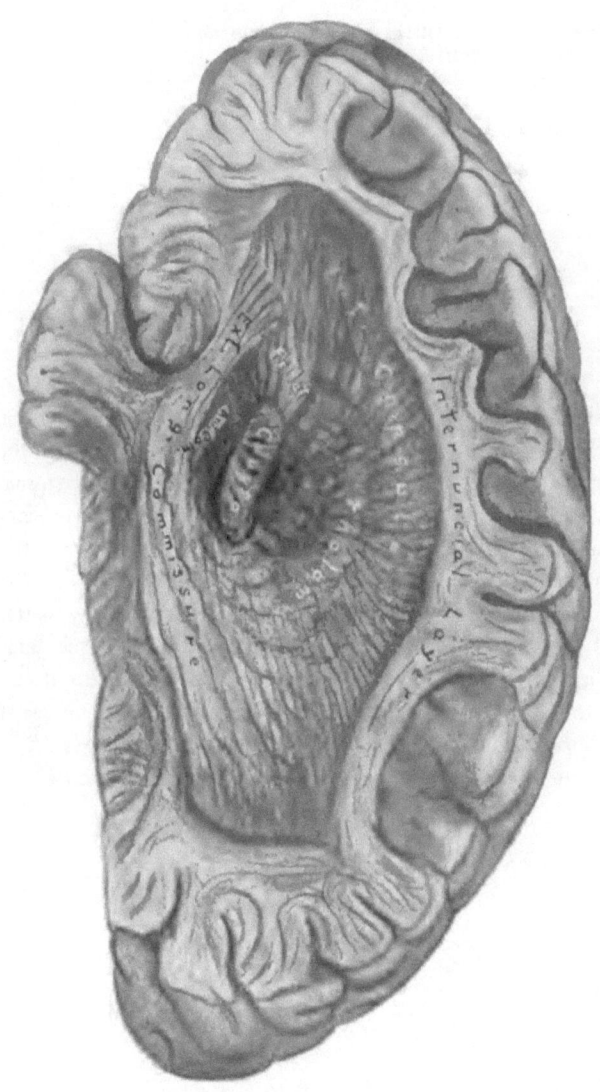

Figure 21.

Figure 22. The external longitudinal commissure has been removed. The anterior commissure of the third ventricle is fully exposed. Behind it are the divided ends of the crusta and the fillet. Above is seen a portion of the external surface of the caudate nucleus, from which the internal capsule has been removed. The thalamus consists of concentric laminæ, between which are layers of grey matter. At the bottom of the cavity, made by washing out the grey matter of the thalamus, is seen the red nucleus of the tegmentum. The fibres of the corpus callosum are distinctly seen to decussate with those of the internal capsule. The laminæ of which each is composed overlie each other, and break off in pointed plates, as seen in the posterior part of the figure.

Figure 22.

Figure 23—Is an intrahemispherical view of the internal surface of a hemisphere, in which the fissures are seen as projections upon the surface, and the convolutions as depressions. The internal capsular fibres are entirely removed, and the laminæ of the corpus callosum are seen projecting from the surface. Above the corpus callosum are the longitudinal fibres of the gyrus fornicatus, called the superior longitudinal commissure, which is seen from above in figure 3. Beneath the corpus callosum is a portion of the lining membrane of the lateral ventricle which covers the ventricular surface of the caudate nucleus. The ventricle is opened behind the thalamus, exposing a portion of the fornix and the hippocampi. The concentric layers which compose the thalamus are well seen in the figure, as well as the relative positions of this body, the fillet, crusta and the anterior commissure of the third ventricle.

Figure 23.

CEREBRAL LOCALIZATION.

The observation of the results of disease and accidents in man, and the experiments of Ferrier, Horsley and others, upon the brains of the lower animals, have established the fact, that certain parts of the hemispherical ganglion are endowed with special functions. The areas that receive the impressions from certain sensory nerves, and that control the movements of certain parts of the body, are definitely mapped upon the surface of the brain, and experience has demonstrated that these discoveries are worthy of confidence.

It is not to be supposed, however, that a science which is so recent in its origin has reached a stage of perfection which is absolutely reliable, but sufficient is already proven to warrant the belief that more definite knowledge is at hand.

It has long been known that injuries and inflammation involving the central part of the surface of the brain are productive of convulsions; that an injury to the third frontal convolution of the left side, (Broca's convolution,) destroys the ability to speak; that disturbances of the intellect occur in affections of the anterior lobes of the brain; and that memory suffers from disease of the occipital lobes. These general observations are confirmed by more recent and careful investigation. The areas devoted to the general functions of the mind, viz.: judgment, memory, and motion, are now more definitely described, and subdivided into parts to which specific functions are attributed. The motor area is situated about the fissure of Rolando, viz.: in the posterior portion of the frontal convolutions, the anterior and posterior central convolutions, and in the superior parietal lobule. This area is divided into smaller portions, which are associated with certain muscular movements, as indicated upon Fig.

Motor area of the brain.

24. Common sensation and muscular sense are situated above and behind the posterior extremity of the fissures of Sylvius; hearing and the memory of words are located in the upper temporal convolutions; smell and taste, in the apex and under surface of the temporal lobe; and vision and the sight memory of words, in the occipital lobes and the cuneus. For more exact information the reader is referred to the works of Ferrier and Horsley. *Area of common sensation and muscular sense. Hearing and word memories. Smell and taste. Vision and sight memories.*

The above observations are intended to direct the attention to the importance of the study of the anatomical relations between the surface of the brain and the surface of the head, and also of the deeper parts of the brain to its external surface. This is the more important, because the walls of the cavity which contain the brain are rigid, requiring the surgeon to direct his operations upon this organ as near the seat of the lesion as possible. We have seen that the structures within the brain are more or less important for the immediate support of life and mental integrity, therefore, it is quite essential that the surgeon should know the situation and direction to the parts lying beneath the point exposed by the trephine, and the safe direction in which he may venture to puncture a ventricle, open an abcess, or use the knife in the removal of a tumor. The cranial cavity is a very important surgical space, and the invader should be better acquainted with its anatomy than with that of any other region, on account of the important functions of its contents. It is not surprising, in the present state of our knowledge of the anatomical relations of the parts of the brain, to learn of recoveries after great violence has been done to this delicate organ, and at other times, that death has resulted from a mere puncture. *Cranial cavity a surgical space.*

For certain reasons, exact localization is an impossibility, though practically this may be obtained within approximate limits. The brain may be exposed, if not in the precise locality sought after, yet near to the situation, which may be recognized by appearance or electrical test. It is of little consequence that a larger portion of the skull is afterward *Reasons why exact localization is impossible.*

removed, in order to reach the desired point. Experience in the examination of different skulls and heads is of great assistance in determining the situation of the prominent fissures and convolutions beneath the surface, enabling the surgeon to place his finger over them without effort, or the assistance of instruments. The variations in different subjects are due: first, to the fact before mentioned, that no two brains are identically marked, nor are the opposite sides of the same brain symmetrical; second, there is frequently a disproportion between the cerebrum and the cerebellum; when the latter organ is large and deep in its vertical diameter, it has the effect of elevating the posterior part of the cerebrum and the vertex of the skull, thereby deranging the lines drawn upon the head as guides to the parts beneath; third, what is true of the brain may also be said of the skull, that parts which are prominent in some skulls are almost wanting in others; for example, the parietal and frontal eminences, the prominence of the occiput, and the nasion may be obscure, and the superior angles of the frontal and occipital bones, in some instances displaced. Experience and judgment is as essential to the aspirant for success in cerebral surgery, as it is to the pelvic surgeon, who also finds in the pelvic region variations which must influence his judgment in particular cases.

System of localization presented.

In devising, therefore, any system of cerebral localization, it is requisite that we should adopt that method which affords the best means under the circumstances of selecting the most reliable guides. In the method of localization here presented, the instrument required is a common string, which is always at hand. The location of a few prominent points and lines, which are easily found and remembered, is all that is necessary. These are, (see figure 23½): the nasion, or root of the nose; the occiput; the vertex, which is midway between the latter two points; the prominences of the angles of the frontal and the occipital bones, situated in the median line of the head; the former about one inch and a half in front of the vertex, and the latter about midway between the vertex

Prominent points and lines upon the skull.

and the occiput. Other points are: the external angular process of the frontal bone, outer canthus of the eye, inferior angle of the malar bone, external auditory meatus, and the mastoid process of the temporal bone.

Beneath the median line of the skull, from the nasion to the occiput, is the superior longitudinal sinus, and the longitudinal fissure separating the hemispheres of the cerebrum. The calloso-marginal fissure emerges from the median fissure a little behind the vertex, and the anterior arm of the parieto-occipital fissure reaches the surface just above the superior angle of the occipital bone. The upper margins of the lobes of the internal surfaces of the hemispheres occupy spaces along the median line: the lingualis and cuneus from the occiput to the upper angle of the occipital bone, the quadratus from the occipital angle nearly to the vertex, the paracentral lobule from the vertex to the angle of the frontal bone, and the marginal convolution from this point to the nasion.

Position of the superior longitudinal sinus, calloso-marginal fissure, and parieto-occipital fissure. Figs. 1 17 20.

Lobes forming the margins of the median fissures. Figs. 1 17 20.

A line drawn from the outer canthus to the center of the parietal eminence covers the fissure of Sylvius, and if continued this line passes through the crown to the opposite auditory meatus; another line drawn from the inferior angle of the malar bone to the vertex indicates the position of the fissure of Rolando.

Lines indicating the fissures of Sylvius and Rolando. Fig. 23½.

The fissure of Sylvius reaches the outer surface of the hemisphere at the intersection of these lines, and divides into a short anterior arm which is directed upward and forward into the third frontal convolution, and a long or posterior arm which extends from the point of division of the fissure, backward and upward, to the middle of the parietal eminence. The center of the parietal eminence is at the intersection of a line from the vertex to the auditory meatus, with the line marking the fissure of Sylvius.

Situation of the fissure of Sylvius.

The fissure of Rolando begins below, about the middle of the posterior arm of the fissure of Sylvius, and, extending directly upward for a short distance, reaches the before mentioned line from the inferior angle of the malar bone to the vertex, under the upper portion of which it lies. The anter-

Of the fissure of Rolando.

ior and posterior central convolutions bound the margins of the fissure of Rolando. The frontal convolutions lie beneath the frontal bone; the first occupying the space between the frontal eminence and the median line; the third, surrounding the anterior arm of the fissure of Sylvius, lies in the temporal fossa, behind the external angular process. The intervening space is occupied by the two divisions of the second frontal convolution. The temporal convolutions lie beneath the fissure of Sylvius and extend upward and backward parallel to it. The supramarginal, or second parietal convolution, surrounds the posterior extremity of the Sylvian fissure, and occupies the region of the parietal eminence. The first parietal lobule is above the parietal eminence, and is bounded above by the margin of the lobus quadratus. The third parietal convolution, or gyrus angularis, is behind and slightly above the parietal eminence, in front of the lambdoid suture. The occipital convolutions are covered by the occipital bone, and border upon the cuneus and lingualis, with which they are connected.

An important point to be located is situated above and behind the auditory meatus. It is indicated by the intersection of two lines, one drawn from the auditory meatus to the superior occipital angle, and the other from the external angular process of the frontal bone to the occiput. The point of intersection marks the outer extremity of the lateral sinus, and also the outer extremity of the superior border of the petrous portion of the temporal bone, which divides the posterior from the middle fossa of the skull. The lateral sinus is directed backward from this point to the occiput, and is usually lower down upon the left than the right side of the head, on account of the larger size of the left hemisphere of the cerebrum, and the smaller size of the left hemisphere of the cerebellum.

Figure 24 shows the situation of the various sensory and motor centers of the convolutions, according to Horsley; and figure 25 the same after Ferrier. By a comparison of these figures, it will be observed that there is an agreement which is corroborative. Figures 25 and 26 are marked by vertical lines into thirty-six sections, which are one-sixth of an inch apart, and numbered to correspond with those of the figures that follow after.

By the use of the topographical sections, it will be easy to study the relations of the deeper parts to the surface of the brain; and by a comparison of successive sections, the deeper parts may be traced in their relations to each other, and compared with the dissections previously studied. The direction in which a trocar may be thrust with safety, and with the least danger of injury to important structures, can be determined at a glance. An examination of these sections creates a suprise that, in so short a distance as one-sixth of an inch, the configuration of the successive sections present so great a difference in appearance.

The location of the deeper parts of the brain by the aid of topographical sections.

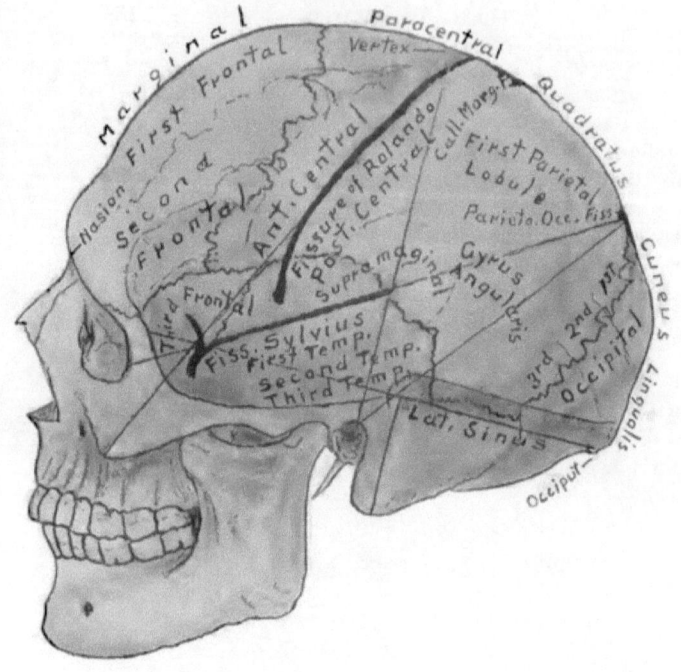

Figure 23½.

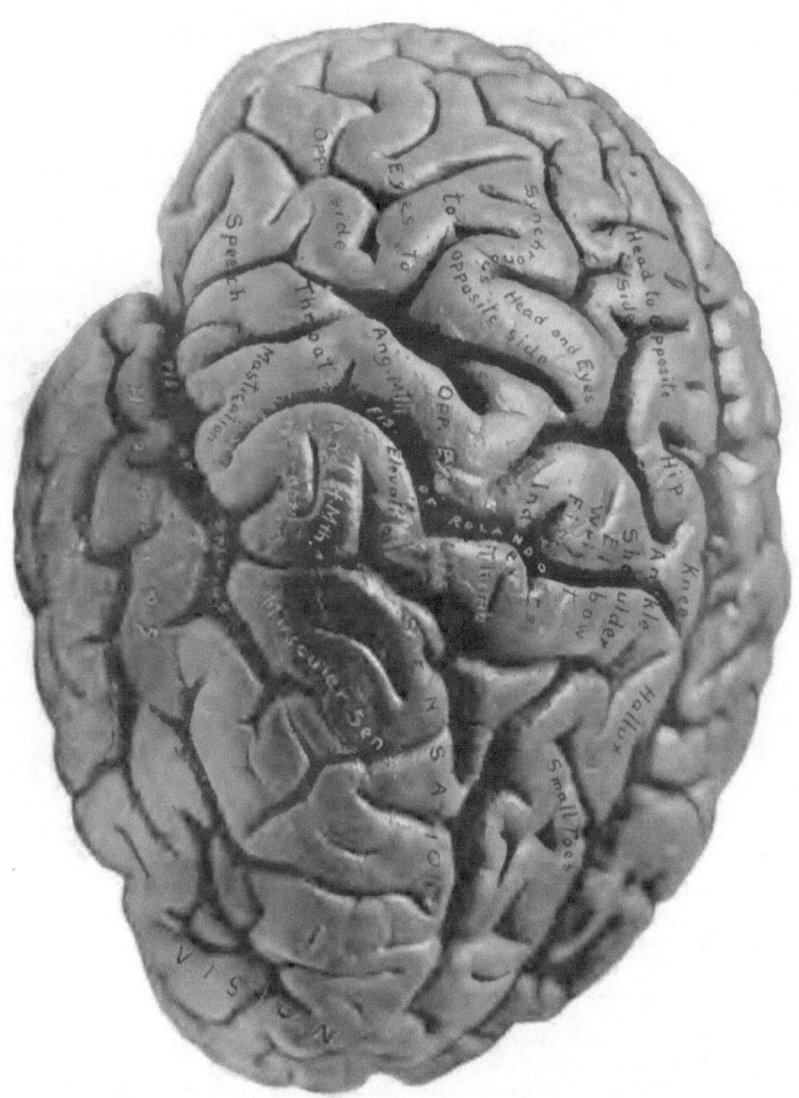

Figure 24.

TOPOGRAPHICAL SECTIONS OF THE BRAIN.

External Surface of the Hemisphere.

LOBES.

F1 F2 F3—Frontal. AC and PC—Anterior and posterior central. T1 T2 T3—Temporal. P1 P2 P3—Parietal. P2—Often called supramarginal. P3—Gyrus angularis. O1 O2 O3—Occipital.

FISSURES.

The fissure of Sylvius is situated above and along the first temporal lobe, and is indicated by a line drawn on the surface of the head from the outer canthus of the eye to the parietal eminence, and if continued the line would pass through the crown to the opposite auditory meatus. The fissure of Rolando is between the lobes AC and PC, and is indicated by a line drawn from the lower angle of the malar bone to the vertex. To locate the lobes and fissures along the longitudinal fissure: the cuneus occupies the space between the occiput and the crown, where the fissure PO reaches the surface; the quadrate lobe extends from the crown to the vertex, where the fissure CM reaches the surface; the paracentral lobule extends from the vertex to the superior angle of the frontal bone; the frontal bone covers the marginal convolution. These directions though not exact are sufficient for practical purposes and are quite as reliable as any of the expensive contrivances in use for locating these parts. The outer portion of the Great Transverse fissure is between the occipital lobes and the cerebellum.

PSYCHOMOTOR CENTERS. (From Ferrier.)

1. Advance of the opposite hind limb as in walking
2, 3, 4. Complex movements of the opposite leg and arm and of the trunk, as in swimming.
a, b, c, d. Individual and combined movements of the fingers and wrist of the opposite hand. Prehensile movements.
5. Extension forward of the opposite arm and hand.
6. Supination and flexion of the opposite forearm
7. Retraction and elevation of the opposite angle of the mouth by means of the zygomatic muscles.
8. Elevation of the alæ nasi and upper lip with depression of the lower lip on the opposite side.
9, 10. Opening of the mouth with (9) protrusion, and (10) retraction of the tongue. *Region of Aphasia.*
11. Retraction of the opposite angle of the mouth, the head turned slightly to one side.
12. Eyes open widely, pupils dilating, and the head and eyes turning towards the opposite side.
13, 14. The eyes moving towards the opposite side with an upward (13), or downward (14) deviation. Pupils generally contracting. (Center of vision.)
15. Pricking up of the opposite ear, head and eyes turning to the opposite side, and pupils dilating largely. (Center of hearing.)

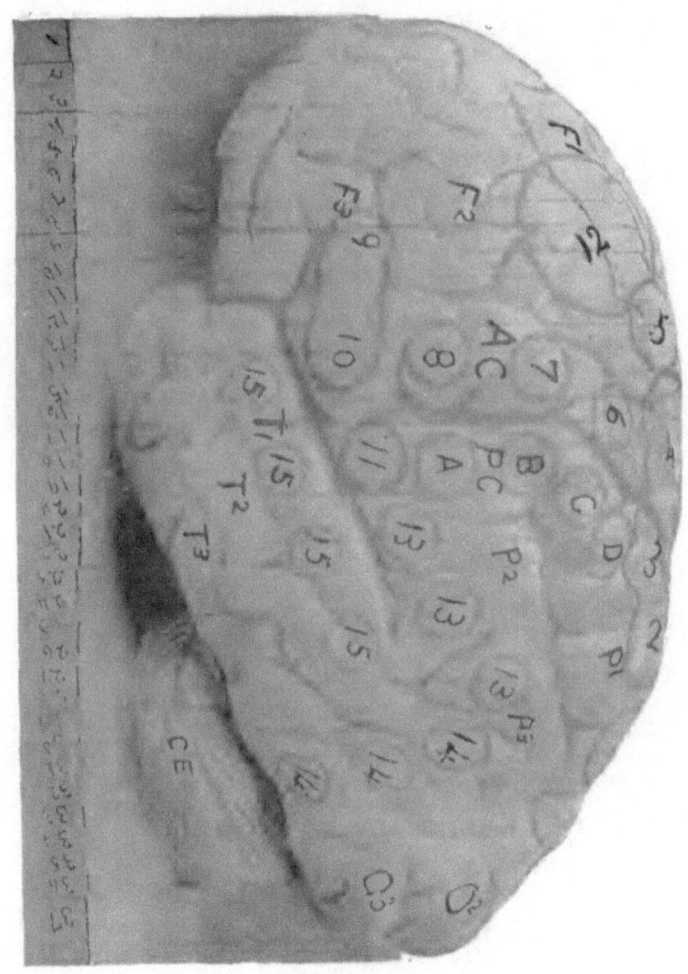

Figure 25.

TOPOGRAPHICAL SECTIONS OF THE BRAIN.

(These sections are cut six to the inch, and the plates are made from photograph of the sections.)

Median Section along the Great Longitudinal Fissure.

LOBES.

L—Lingualis. C—Cuneus. Q—Quadratus. GF—Gyrus fornicatus. PC—Paracentral. MM—Marginal.

FISSURES.

The fissure of Bichat, or the Great Transvers fissure, is between the lingual lobe and the cerebellum. The Calcarine fissure is between L and C. The Parieto-occipital fissure is between C and Q. The Calloso-marginal fissure is bounded by the lobes Q and GF below, and PC and MM above. The fissure called the Ventricle of the Corpus Callosum is between GF and CC.

COMMISSURES, ETC.

CC—Corpus callosum. F—Fornix. SL—Septum lucidum. Beneath F is the intraventricular portion of the Great Transverse fissure, through which passes the velum interpositum from the third to the lateral ventricle, its anterior extremity is called the Foramen of Monro. AC—Anterior commissure. MC—Middle commissure. P—Pineal gland, its anterior pillars extending forwards below the fornix. In front of and beneath P, is the posterior commissure. TN—Corpora quadrigemina —T is on the testes, and N on the nates. Beneath T and N is the iter. CE—Cerebellum. Between CE and T is the valve of Vieussens, beneath which is the cavity of the fourth ventricle. Above and below the valve are the superior and inferior vermiform processes. R—Restiform body, beneath which is the central canal of the cord. DP—Decussation anterior pyramids. N—Pons Varolii. FT—Fillet or lemniscus. Between FT and the fourth ventricle is the fasciculi teretes. PT—Decussation of the processes testes. RN—Crus cerebri, the letters are over the red nucleus. 3—Third nerve. CA—Corpora albicantia. I—Infundibulum and tuber cinereum. O—Optic commissure. Between O and AC is the lamina cinerea, in front of which is the anterior pillar of the the corpus callosum.

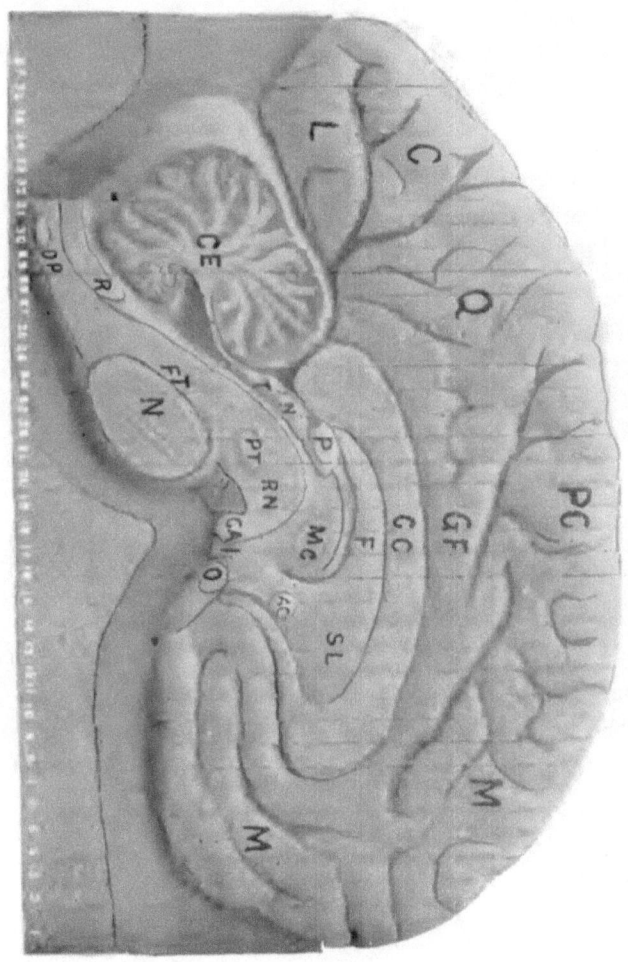

Figure 26.

140 ARCHITECTURE OF THE BRAIN.

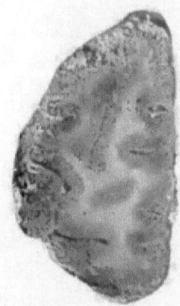

No. 1.

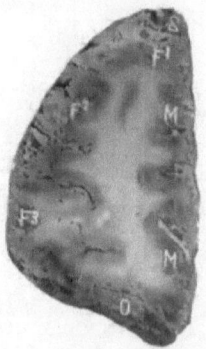

No. 2.

M and M—Marginal conv.
F¹ F² F³—Frontal convolutions.
O—Orbital convolution.

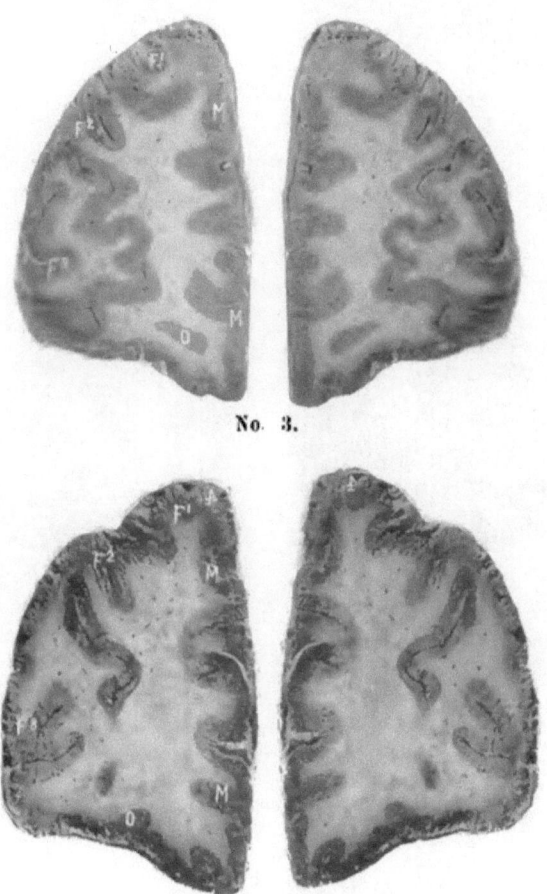

No. 3.

No. 4.

M and M—Marginal conv.
F¹ F² F³—Frontal convolutions.
O—Orbital convolution.

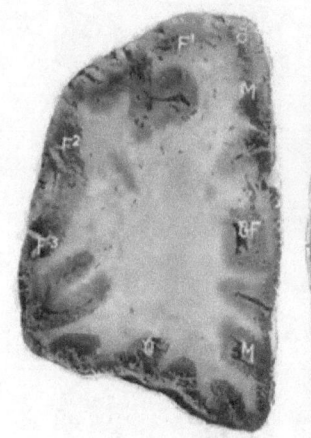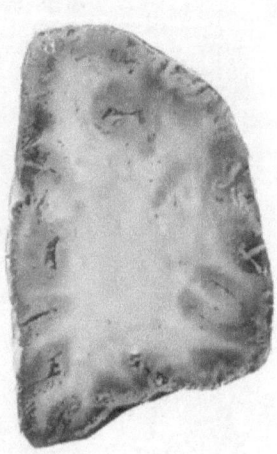

No. 5.

G F—Gyrus fornicatus.
M and M—Marginal conv.
F¹ F² F³—Frontal convolutions.
O—Orbital convolution.

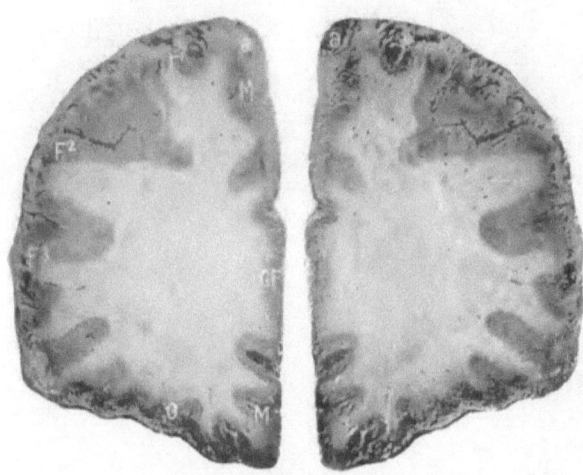

No. 6.

G F—Gyrus fornicatus.
M and M—Marginal conv.
F¹ F² F³—Frontal convolutions.
O—Orbital convolution.

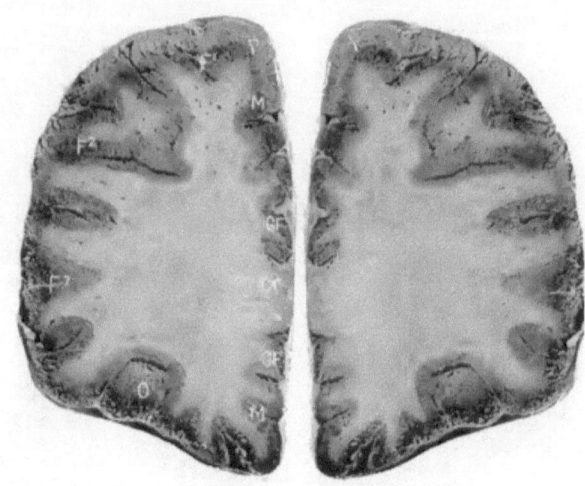

No. 7.

C C—Corpus callosum.
G F—Gyrus fornicatus.
M and M—Marginal conv.
F¹ F² F³—Frontal convolutions.
O—Orbital convolution.

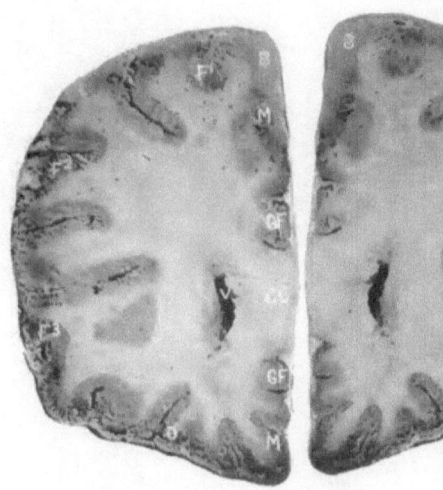

No. 8.

C C - Corpus callosum.
G F—Gyrus fornicatus.
M M—Marginal convolution.
F¹ F² F³—Frontal convolutions.

O—Orbital convolution.
V—Anterior cornu of lateral ventricle.

146   ARCHITECTURE OF THE BRAIN.

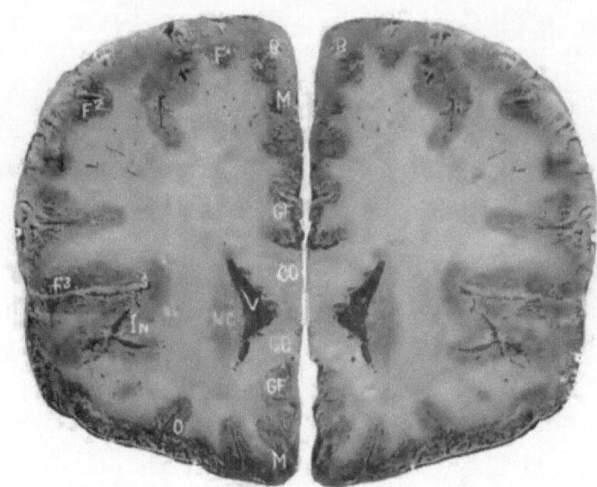

No. 9.

C C - Corpus callosum.
G F—Gyrus fornicatus.
M M—Marginal convolution.
F¹ F² F³ —Frontal convolution.
O—Orbital convolution.

V—Anterior cornu of lateral ventricle.
N C—Nucleus caudatus.
C L—Claustrum.
In—Insula, or island of Riel.
S—Fissure of Sylvius.

TOPOGRAPHICAL SECTIONS OF THE BRAIN.   147

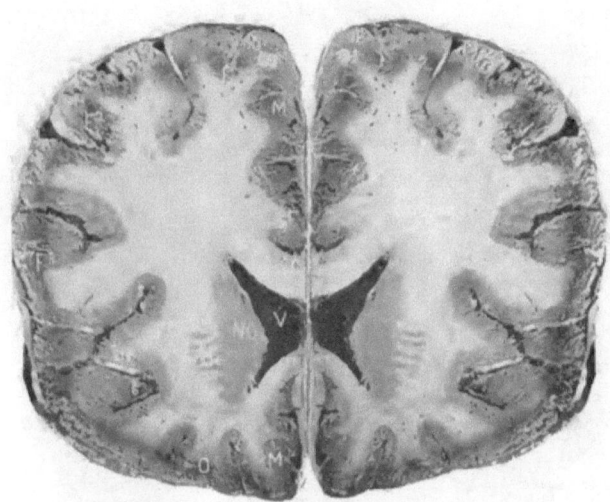

No. 10.

CC and cc—Corpus callosum.
G F—Gyrus fornicatus.
M and M—Marginal conv.
F¹ F² F³—Frontal convolutions.
O—Orbital convolution.

V—Lateral ventricle.
N C—Nucleus caudatus.
I C Internal capsule.
C L—Claustrum.
In—Insula, or island of Riel.
S and S— Fissure of Sylvius.

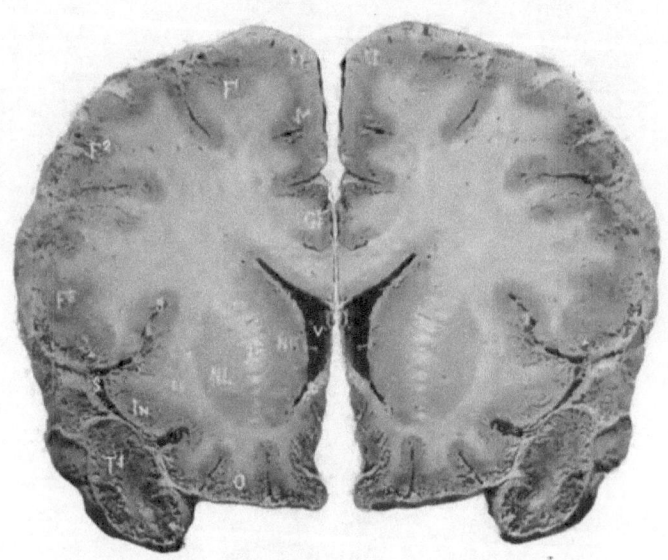

No. 11.

C C and c c —Corpus callosum.
G F—Gyrus fornicatus.
M—Marginal convolution.
F¹ F² F³ —Frontal convolutions.
T¹ —First temporal convolution.
O— Orbital convolution.

V -Lateral ventricle.
N C—Caudate nucleus.
I C—Internal capsule.
N L—Lenticular nucleus.
X—External capsule.
C L—Claustrum.
In— Insula, or island of Riel.
S and S--Fissure of Sylvius.

TOPOGRAPHICAL SECTIONS OF THE BRAIN. 149

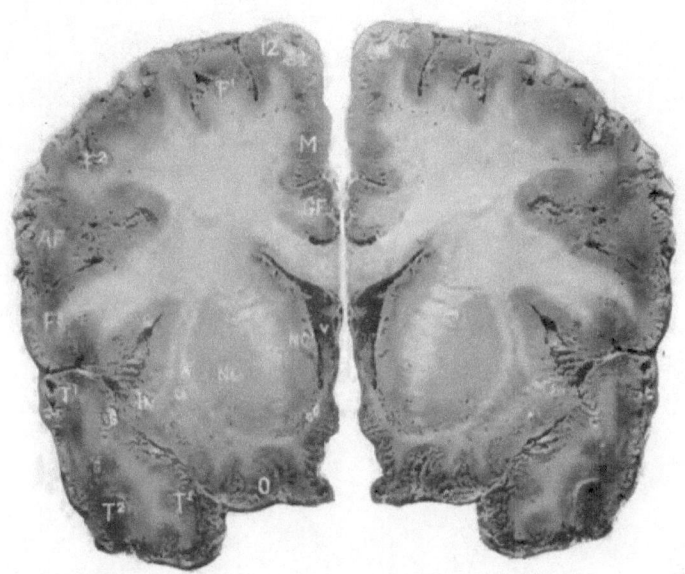

No. 12.

CC and cc—Corpus callosum.
G F—Gyrus fornicatus.
M—Marginal convolution.
F¹ F² F³—Frontal convolutions.
A F—Ascending frontal conv.
T¹ T¹ T²—Temporal convs.
O—Orbital convolution.

V—Lateral ventricle.
N C—Nucleus caudatus.
I C—Internal capsule.
N L—Nucleus lenticularis.
X—External capsule.
C L—Claustrum.
In—Insula, or island of Riel.
S and S—Fissure of Sylvius.

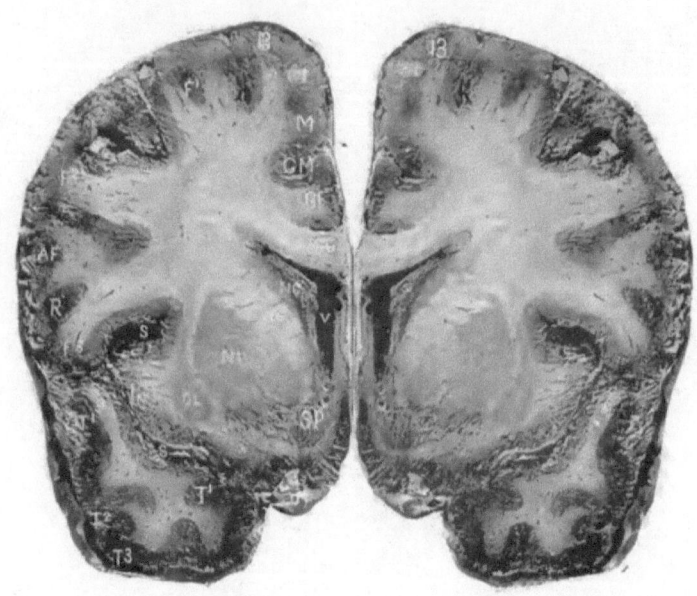

No. 13.

C C—Corpus callosum.
G F—Gyrus fornicatus.
C M—Calloso-marginal fissure
M—Marginal convolution.
F¹ F² F³ Frontal convs.
A F—Ascending frontal conv.
R—Fissure of Rolando.
T¹ T² T³—Temporal convs
O—Optic nerve.
S P—Substantia perforata anterior.

V—Lateral ventricle.
N C—Nucleus caudatus.
I C—Internal capsule.
N L—Nucleus lenticularis.
X—External capsule.
C L—Claustrum.
In—Insula, or Island of Riel.
S and S—Fissure of Sylvius.

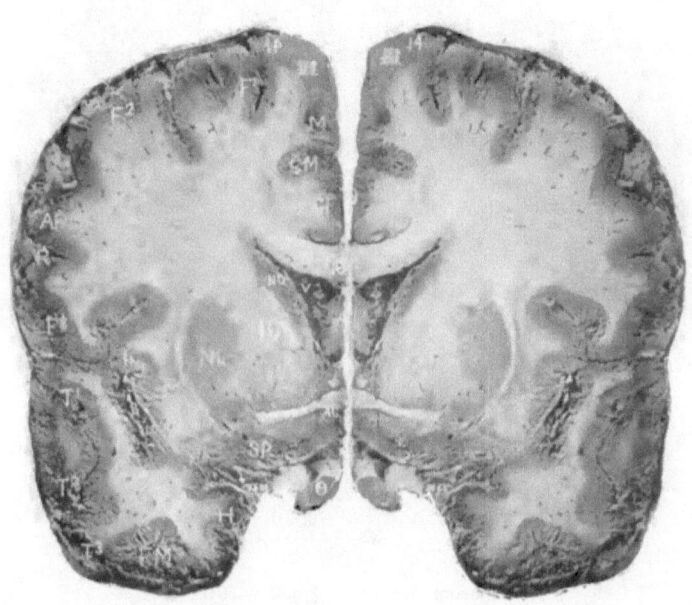

No. 14.

C C—Corpus callosum.
G F—Gyrus fornicatus.
C M—Calloso-marginal fissure
M—Marginal convolution.
F¹ F² F³—Frontal convs.
A F—Ascending frontal or central convolution.
R—Fissure of Rolando.
T¹ T² T³—Temporal convs
F M—Fusiform convolution
H—Hippocampal conv.
S P—Substantia perforata.

O—Optic nerve.
L C—Lamina cinerea.
A C—Anterior commissure.
F—Anterior pillar fornix.
V—Lateral ventricle.
N C—Caudate nucleus.
I C—Internal capsule.
N L—Lenticular nucleus.
X—External capsule.
C L—Claustrum.
In—Insula, or island of Riel.
S and S—Fissure of Sylvius.

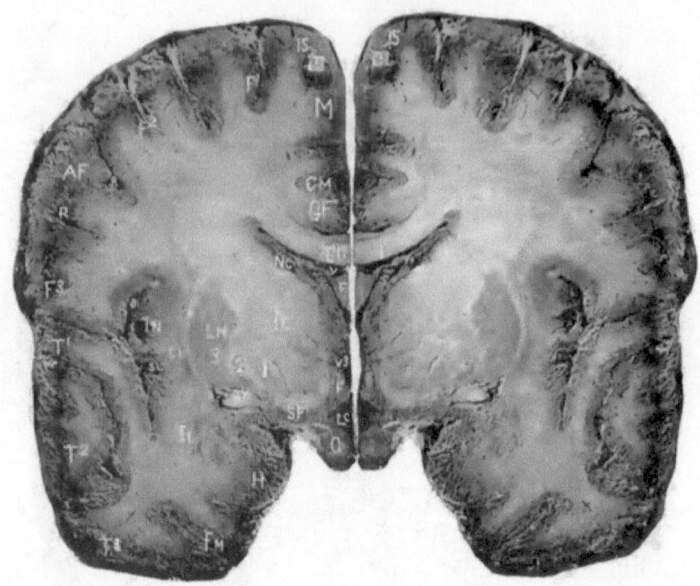

No. 15.

C C—Corpus callosum.
G F—Gyrus fornicatus.
C M—Calloso-marginal fissure
M—Marginal convolution.
$F^1 F^2 F^3$—Frontal convs.
A F—Ascending frontal or central convolution.
R R—Fissure Rolando.
$T^1 T^2 T^3$—Temporal convs.
F M—Fusiform convolution.
H—Hippocampal convolution
S P—Substantia perforata.
O—Optic commissure.
Lc—Lamina cinerea.

F and F—Fornix.
V^3—Third ventricle.
V—Lateral ventricle.
N C—Caudate nucleus.
I C—Internal capsule.
L N—Lenticular nucleus. 1, 2, 3, its three divisions.
A C—Divided outer extremity of the anterior commissure.
X—External capsule.
C L—Claustrum.
In—Insula.
S and S—Fissure Sylvius.

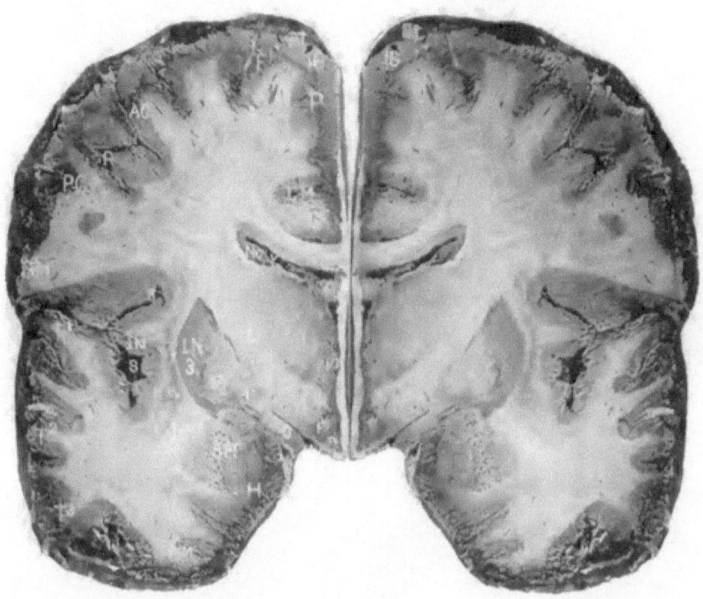

No. 16.

C C—Corpus callosum.
G F—Gyrus fornicatus.
C M—Calloso-marginal fissure
P—Paracentral lobule.
F^1—First frontal conv.
A C—Anterior central or ascending frontal.
P C—Posterior central or ascending parietal.
R—Fissure of Rolando.
S M—Supramarginal or second parietal lobule.
T^1 T^2 T^3—Temporal convs.
F M—Fusiform convolution.
H—Hippocampal conv.

S P P—Substantia perforata posterior.
O—Optic tract.
T C—Tuber cinereum.
F and F—Fornix.
V^3—Third ventricle.
V—Lateral ventricle.
N C—Tail of caudate nucleus.
T H—Thalamus opticus.
IC and IC—Internal capsule.
LN $^{1\,2\,3}$—Lenticular nucleus.
X—External capsule.
C L—Claustrum.
In—Insula, or island of Riel.
S and S—Fissure of Sylvius.

154   ARCHITECTURE OF THE BRAIN.

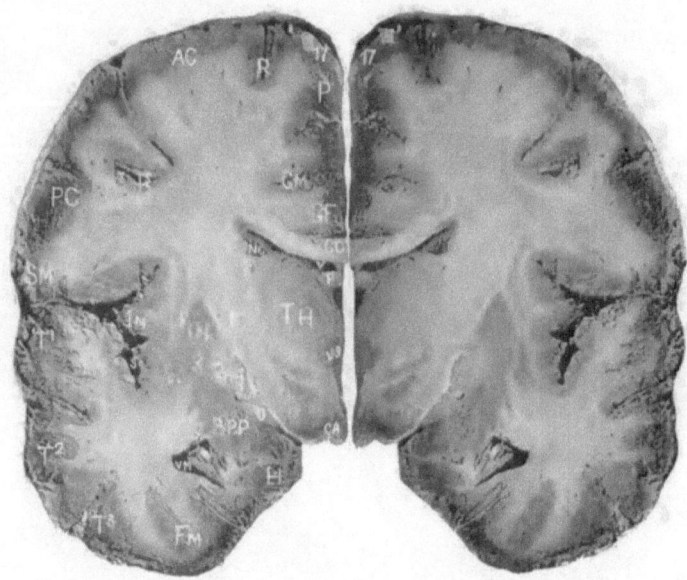

No. 17.

C C—Corpus callosum.
G F—Gyrus fornicatus.
C M—Calloso-marginal fissure
P—Paracentral lobe.
R and R— Fissure of Rolando.
A C and P C—Anterior and posterior central convs.
S M—Supramarginal or second parietal convolution.
T¹ T² T³ —Temporal convs.
F M—Fusiform convolution.
H—Hippocampal conv.
V M - Middle cornu of lateral vent. and pes hippocampi.
S P P—Substantia perforata posterior.
O—Optic tract.

C A—Corpus albicantia.
V³—Third ventricle. The letter rests upon the grey matter of its wall opposite the middle commissure.
F—Fornix.
V—Lateral ventricle.
N C—Tail of caudate nucleus.
T H—Thalamus.
IC and IC—Internal capsule.
LN¹ ² ³ —Lenticular body; its three divisions.
X—External capsule.
C L—Claustrum.
In—Insula, or island of Riel.
S and S—Fissure of Sylvius.

TOPOGRAPHICAL SECTIONS OF THE BRAIN.

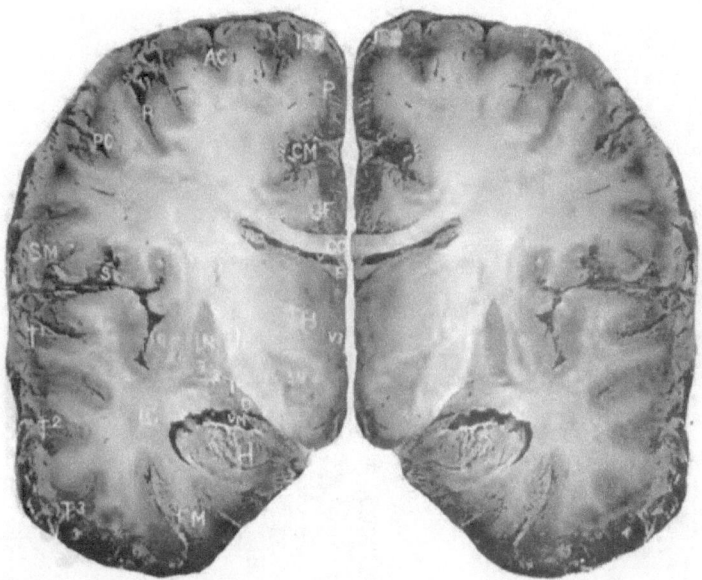

No. 18.

C C—Corpus callosum.
G F—Gyrus fornicatus.
C M—Colloso-marginal fissure
P—Paracentral lobule.
A C and P C—Anterior and posterior central convs.
R—Fissure of Rolando.
S M—Supramarg. conv.
$T^1\ T^2\ T^3$—Temporo sphenoidal, or temporal convs.
F M—Fusiform convolution.
H—Hippocampal. The letter rests upon the cornu Ammonis.
V M—Middle cornu lat. vent.
O—Optic tract.
S N—Substantia nigra of crus cerebri, and external to SN are the fibres of the crusta passing upward to the internal capsule.
V^3—Wall of third ventricle.
F—Fornix.
V—Lateral ventricle.
N C and N C—Tail of caudate body; the lower NC is external to O and is the lower extremity of the tail of the caudate body, situated in the roof of the middle cornu of the lateral ventricle.
IC and IC—Internal capsule; the lower IC are the fibres of the capsule distributed to the temporal lobes.
$L\ N^{1\ 2\ 3}$—Lenticular body.
X—External capsule.
C L—Claustrum.
In—Insula, or island of Riel.
S—Fissure of Sylvius.

156　　ARCHITECTURE OF THE BRAIN.

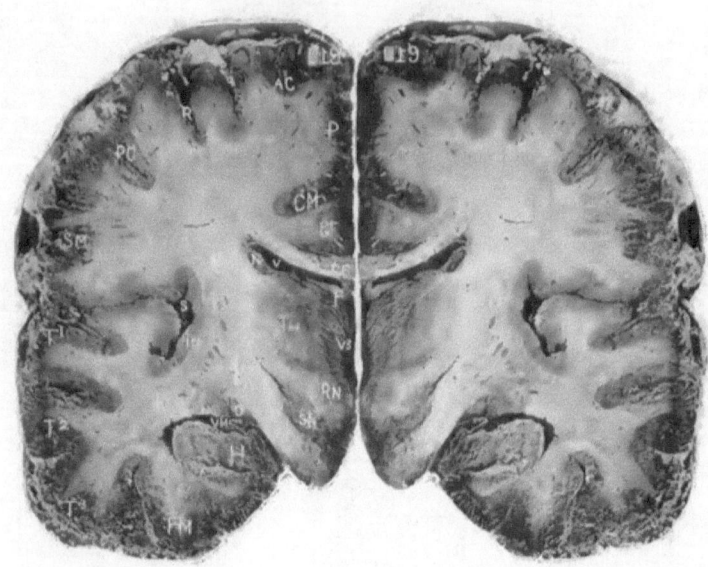

No. 19.

C C—Corpus callosum.
G F—Gyrus fornicatus.
C M—Calloso-marginal fissure
P—Paracentral lobe.
A C and P C—Anterior and posterior central convs.
R—Fissure of Rolando.
S M—Supra marginal conv.
T¹ T² T³—Temporal convs.
F M—Fusiform lobe.
H—Hippocampal.
V M—Middle cornu lateral ventricle.
O—Optic tract.
S N—Substantia nigra, or locus niger. The crusta is external to SN, and is continuous with IC.

R N—Red nucleus, running out from which is the fillet.
V³—Wall of third ventticle.
F—Fornix.
V—Lateral ventricle.
NC and NC—Caudate nucleus.
TH.—Thalamus.
IC, IC, IC—Internal capsule. The row of dark spots between the upper and lower IC is the posterior extremity of the lenticular nucleus.
In—Insula, or island of Riel.
S—Fissure of Sylvius. A trace of the claustrum can be seen in this section.

TOPOGRAPHICAL SECTIONS OF THE BRAIN.

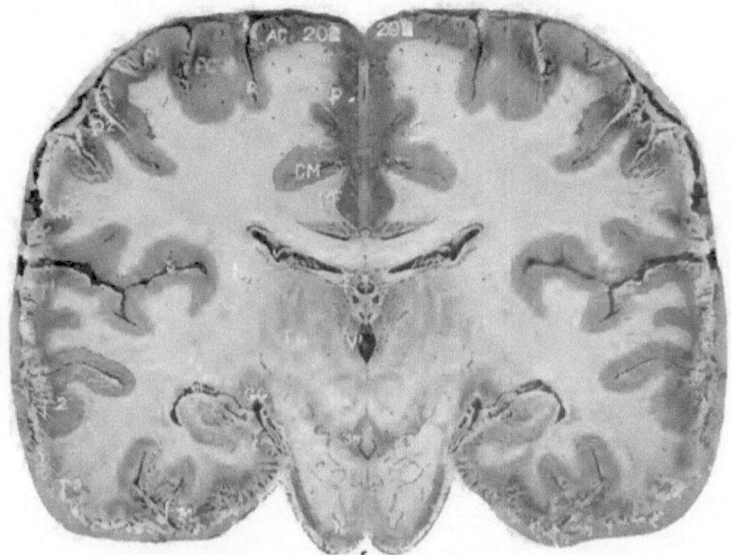

No. 20.

C C—Corpus collosum.
G F—Gyrus fornicatus.
C M—Calloso-marginal fissure
P—Paracentral lobe.
A C and P C—Anterior and posterior central convs.
R—Fissure of Rolando.
P² —Paracentral lobe.
S M—Supramarginal conv.
T¹ T² T³—Temporal, or tempero sphenoidal convs.
F M—Fusiform.
H — Hippocampal convolution, or cornu Ammonis, the projection of which into the middle cornu of the lateral ventricle forms the hippocampus major.
V M—Middle cornu.

G G—Corpus geniculata, external and internal, which are the termination of the two roots of the optic tract.
N—Pons Varolii.
S N—Substantia nigra.
R N—Red nucleus, which is connected by a grey commissure with another round grey mass above, and internal to the letters TH.
V³—Third ventricle.
F—Fornix.
V—Lateral ventricle.
NC and NC—Caudate nucleus.
T H—Thalamus, displaying a peculiar radiation of fibres.
I C and I C—Internal capsule at its genu.
S—Fissure of Sylvius.

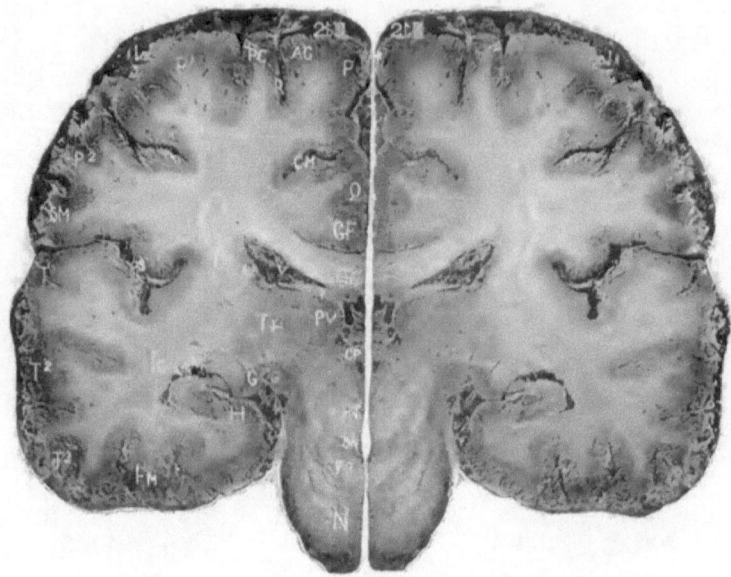

No. 21.

C C—Corpus callosum.
G F—Gyrus fornicatus.
Q—Quadratus or precuneus lobe.
C M—Calloso-marginal fissure
P. and A C—Union of paracentral and anterior central convolutions.
R—Fissure of Rolando.
P C—Poterior central or ascending parietal conv.
P¹—First parietal lobule.
P² and S M—Second parietal lobule or supra marginal convolution.
T¹ T² T³—Temporal convs.
F M—Fusiform convolution.
H—Hippocampal lobe.
V M—Middle cornu and hippocampus major; and directed toward G G is the corpus fimbriatum or edge of the fornix, forming the lower boundary of the fissure of Bichat, through which the velum enters the lateral ventricle.
G G—Corpora geniculata, continuous below with the optic tract and behind with the pulvinar.
P—Pulvinar.
N—Pons.
F T—Fillet tract.
S N—Substantia nigra.
P T—Processus testes.
C P—Posterior commissure.
F—Fornix.
V—Lateral ventricle.
NC and NC—Caudate nucleus.
I C and I C—Internal capsule.
S—Fissure Sylvius.

TOPOGRAPHICAL SECTIONS OF THE BRAIN.

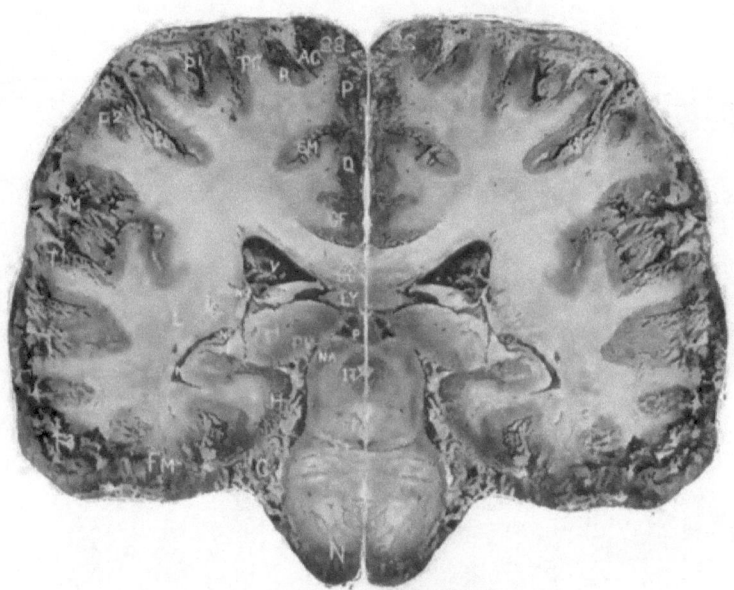

No. 22.

C C—Corpus callosum.
G F—Gyrus fornicatus.
Q—Lobus quadratus.
C M—Calloso-marginal fissure
P and AC—Paracentral and ascending frontal.
R—Upper end of fissure of Rolando.
P C—Posterior central conv.
P¹—First parietal lobule.
P² and S M—Second parietal lobe.
T¹ T² T³—Temporal convs.
F M—Fusiform conv.
H—Hippocampal conv. at its junction with the lingualis and gyrus fornicatus.
N—Pons Varolii.
F T—Fillet or lemniscus.
P T—Processus e cerebello ad testes at its decussation.

I T—Iter.
N A—Nates.
P—Pineal gland.
Ly—Lyra.
P V—Pulvinar, or optic centre of the thalamus.
T H—Thalamus.
F and F—Fornix, winding around the posterior part of the thalamus.
V—Lateral ventricle.
NC and NC—Genu of the tail of the caudate nucleus.
V M—Middle cornu of the lateral ventricle.
I C—Posterior fibres of the Internal capsule.
L—External longitudinal commissure.
S—Fissure of Sylvius.

160   ARCHITECTURE OF THE BRAIN.

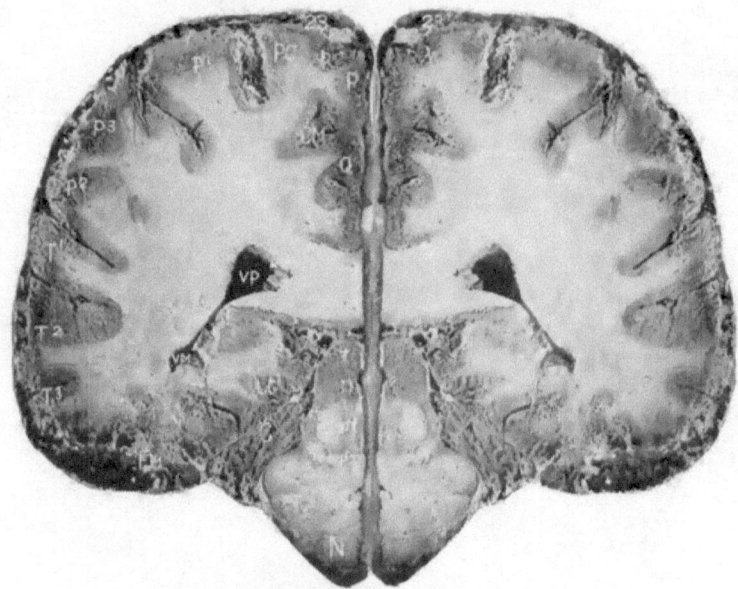

No. 23.

C C—Corpus callosum.
G F—Gyrus fornicatus.
Q—Lobus quadratus.
C M—Calloso-marginal fissure
P—Paracentral lobule.
R—Fissure of Rolando.
P C—Ascending parietal or posterior central conv.
P¹—First parietal lobule.
P³—Third parietal lobule or gyrus angularis.
P²—Second parietal lobule.
T¹ T² T³—Temporal convs.
F M—Fusiform lobe.
L G—Lobus lingualis.
C—Cerebellum.
N—Pons.
F T—Fillet or lemniscus.

P T—Processus e cerebello ad testes.
I T—Iter e tertio ad quartum ventriculum.
T—Testes.
V P—Posterior cornu of the lateral ventricle.
V M—Middle cornu of the lateral ventricle.
I C Posterior distribution of fibres of the internal capsule.
L—External longitudinal commissure, separated from the sub-convolutional layer of the corona radiata by a delicate arched white line.

TOPOGRAPHICAL SECTIONS OF THE BRAIN.

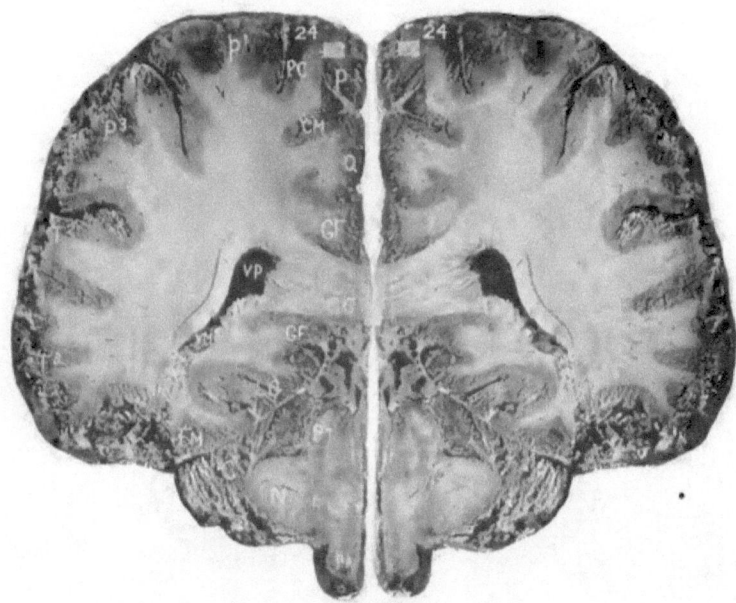

No. 24.

C C—Corpus callosum.
G F and G F—Gyrus fornicatus
Q—Lobus quadratus or precuneus.
C M—Calloso-marginal fissure
P and PC—Junction of paracentral and post. central convolutions.
P¹—First parietal lobe.
P³—Gyrus angularis.
T² T³—Temporal convs.
F M—Fusiform lobe.
L G—Lingualis.
VP and VM—Junction of the posterior with the middle cornu of the lateral ventricle
L—Ext. longitudinal comm.
I C—Internal capsule.
P T—Processes testes.
N—Lateral fibres of pons.

R—Restiform fibres of medulla oblongata.
F T—Fillet.
P R—Anterior pyramids of the medulla.

The composition of the corona radiata is well seen in this plate. The fibres of the corpus callosum are internal to the ventricle; the decussation between its fibres and the internal capsule is above, while the outer wall of the ventricle is formed by the internal capsule, separated from L by a dark line. External to L is the sub-convolutional white fibres, or internuncial layer of the corona radiata.

162 ARCHITECTURE OF THE BRAIN.

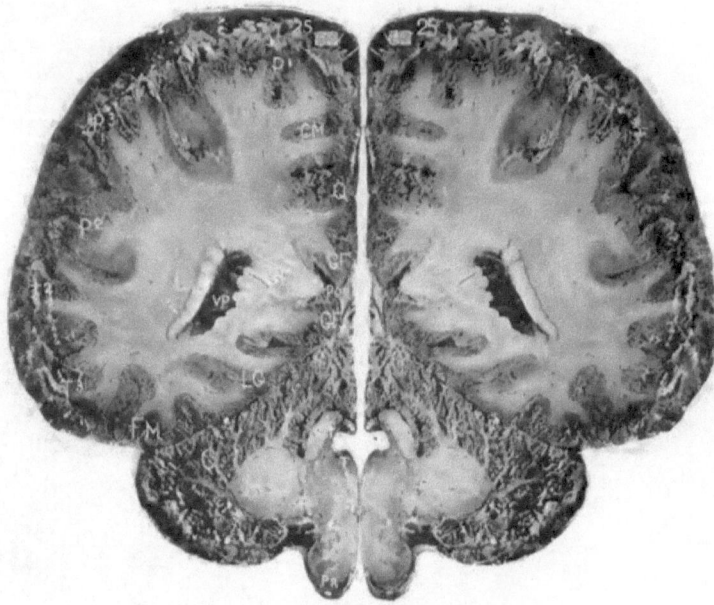

No. 25.

L —External longitudinal commissure.
V P—Posterior cornu. Between L and V P are fibres of the internal capsule directed backward to the cuneus and occipital lobes, containing optic fibres.
C C—Posterior fibres of the corpus collosum.
P O—Commencement of the parieto-occipital fissure.
G F and G F— Gyrus fornicatus
Q—Lobus quadratus.
C M — Calloso-marginal fissure
P¹ — First parietal lobule.
P³ —Lobus angularis.
P² —Supra-marginal or second parietal lobule.

T² T³ —Temporal convolutions.
F M— Fusiform convolution.
L G - Lingualis.
C—Cerebellum.
N— Lateral fibres of pons.
R—Restiform body.
O —Olivary body.
P R—Anterior pyramid.
F T—Fillet.
T—Fasciculus teretes.
P T— Processes testes and fourth ventricle. The valve of Vieussens forms the roof of the fourth ventricle by uniting the upper part of one processus with the other.

TOPOGRAPHICAL SECTIONS OF THE BRAIN.

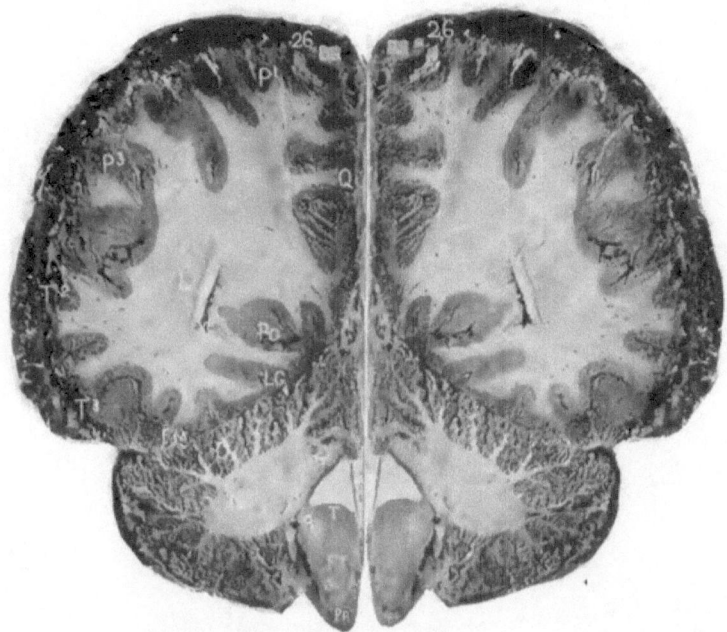

No. 26.

L—External longitudinal commissure and internal capsule.
V P—Posterior cornu.
P O—Parieto-occipital fissure.
Q—Lobus quadratus.
P¹—First parietal conv.
P³—Angular lobe.
T² T³—Temporal convs.
F M—Fusiform convolution.
L G—Lingualis convolution.
C—Cerebellum.
N—Pons, lateral fibres.
P T—Processus testes.
R—Restiform body.
T—Fasciculus teretes.
F T—Fillet or lemniscus.
O—Olivary body.
P R—Anterior pyramids of medulla.

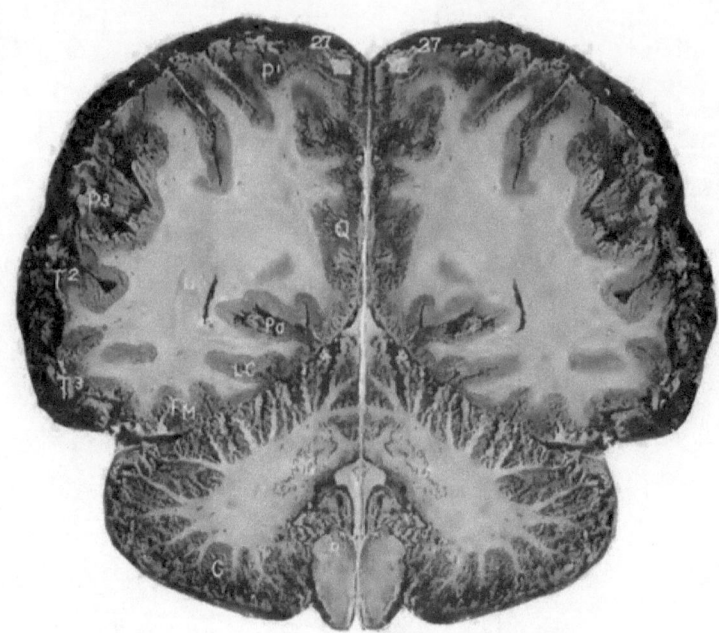

No. 27.

L—External longitudinal commissure.
VP—Posterior cornu and internal capsular fibres.
PO—Parieto-occipital fissure.
Q—Lobus quadratus.
P¹ P³—Parietal lobes.
T² T³—Temporal convolutions.
F M—Fusiform convolution.
L G—Lingualis.
C—Cerebellum.
D—Corpus dentatum.
R—Restiform body.

The olivary body and anterior pyramids are seen beneath R.

TOPOGRAPHICAL SECTIONS OF THE BRAIN.

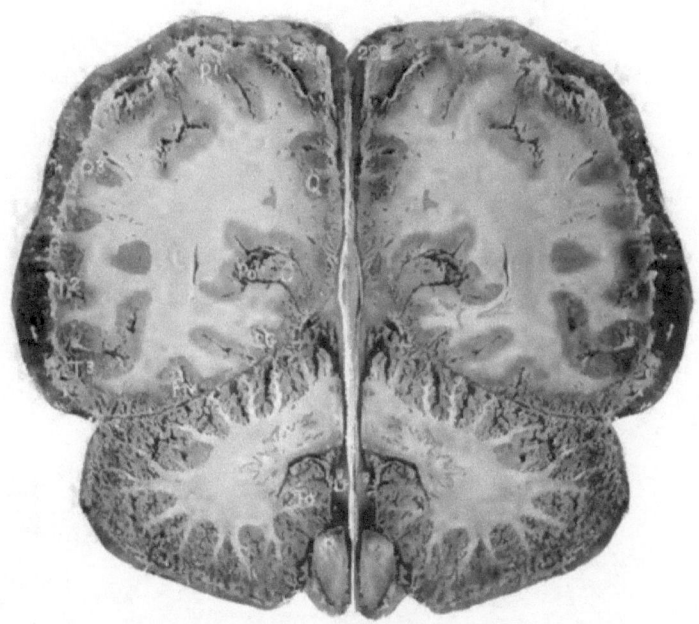

No. 28.

L—External longitudinal commissure.
Posterior fibres of the internal capsule, posterior cornu.
Po—Parieto-occipital fissure.
C—Apex of cuneus.
Q—Quadratus or precuneus.

P¹ P²—Parietal lobes.
T² T³—Temporal lobes.
F fl—Fusiform convolution.
L G—Lingualis.
D—Corpus dentatum.
To—Tonsil.
U—Uvula.
R—Restiform body.

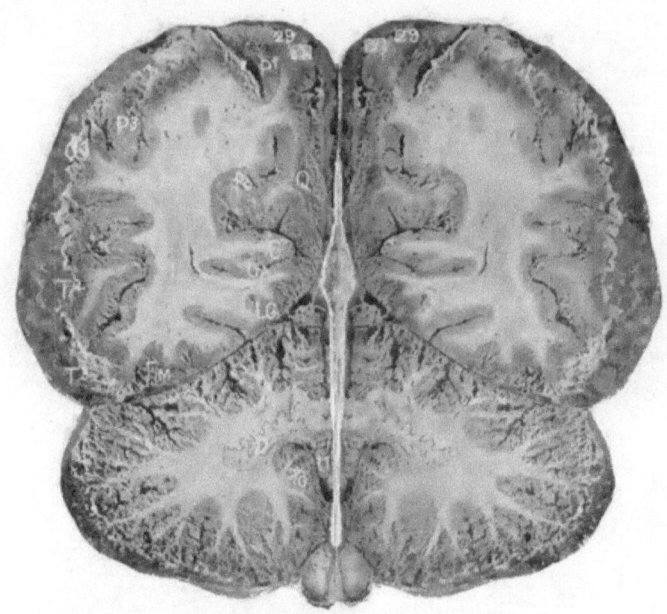

No. 29.

L—External longitudinal commissure, internal capsule, cornu lateral ventricle.
C A—Calcarine fissure.
C—Cuneus lobe.
Po—Parieto-occipital fissure.
Q—Quadratus lobe.
$P^1 P^3$—Parietal lobes.
$T^2 T^3$—Temporal lobes.
F M—Fusiform lobe.
L G—Lingual lobe.
D—Corpus dentatum.
To—Tonsil.
U—Uvula.

TOPOGRAPHICAL SECTIONS OF THE BRAIN. 167

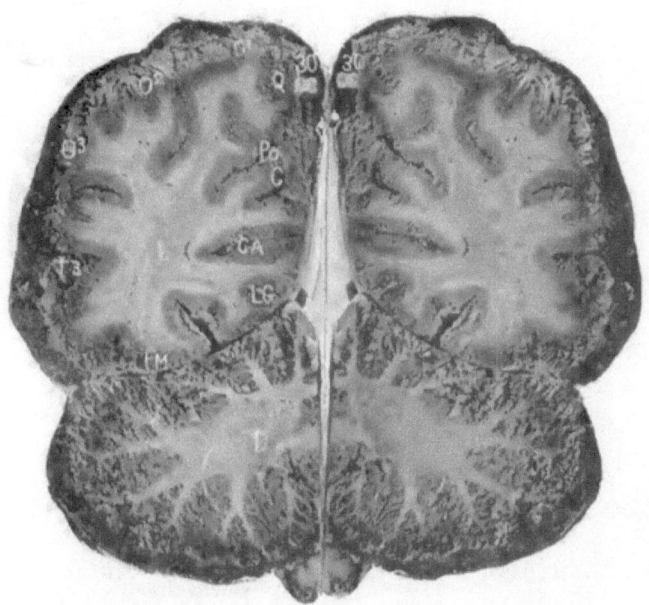

No. 30.

L.—External longitudinal commissure, extremity of posterior cornu.
C A—Calcarine fissure.
C—Cuneus.
Po—Parieto-occipital fissure.
Q—Precuneus or quadratus.
O^1 O^2 O^3 — Occipital lobes.
T^3 —Third temporal conv.
F M—Fusiform.
L G—Lingualis.
D—Dentate body.

168   ARCHITECTURE OF THE BRAIN.

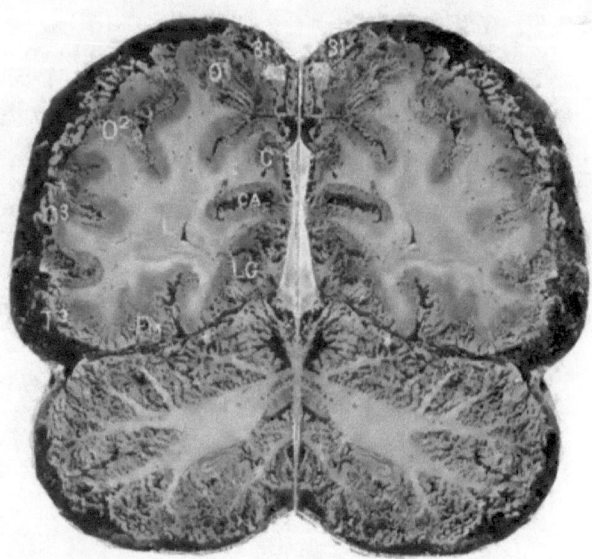

No. 31.

L—External longitudinal commissure, posterior cornu.
C A—Calcarine fissure.
C—Cuneus.

O^1 O^2 O^3—Occipital lobes.
T^3—Third temporal
F M—Fusiform.
L G—Lingualis.

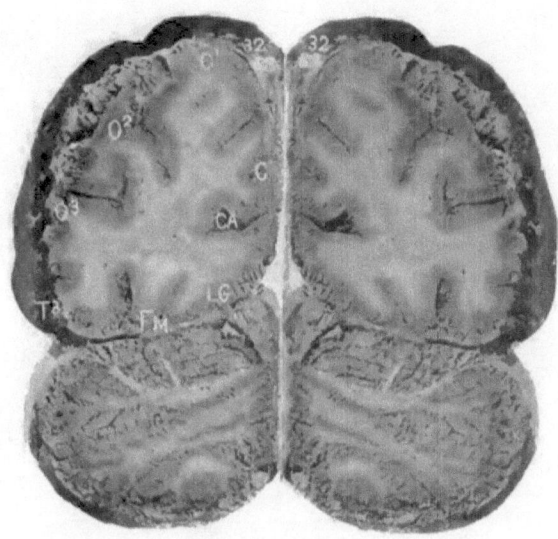

No. 32.

C A—Calcarine fissure.
C—Cuneus.
O¹ O² O³—Occipital lobes.

T³—Third temporal.
F M—Fusiform.
L G—Lingualis.

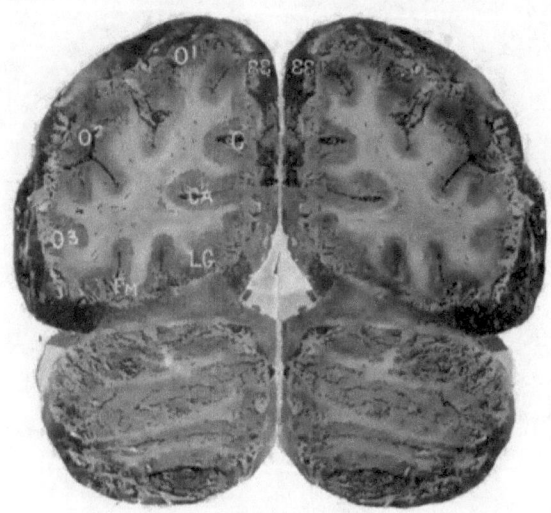

No. 33.

C A—Calcarine fissure.
C—Cuneus.
O¹ O² O³—Occipital convs.
F M—Fusiform.
L G—Lingualis.

TOPOGRAPHICAL SECTIONS OF THE BRAIN. 171

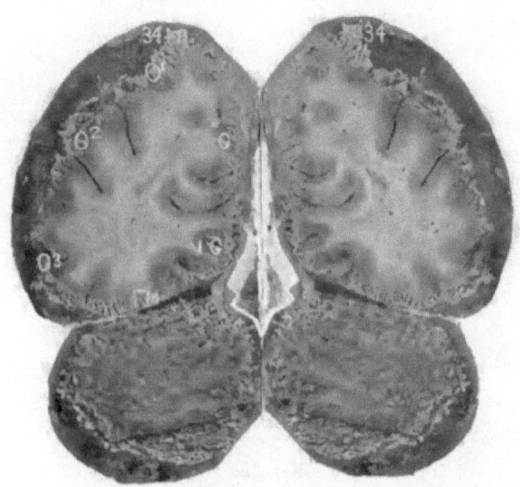

No. 34.

C—Cuneus.
$O^1\ O^2\ O^3$ – Occipital convs.
F M—Fusiform.
L G—Lingualis.

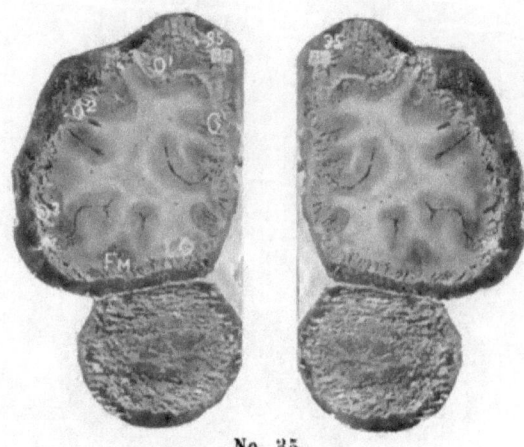

No. 35.

C—Cuneus.
O¹ O² O³ — Occipital convs.
F M—Fusiform.
L G—Lingualis.

TOPOGRAPHICAL SECTIONS OF THE BRAIN. 173

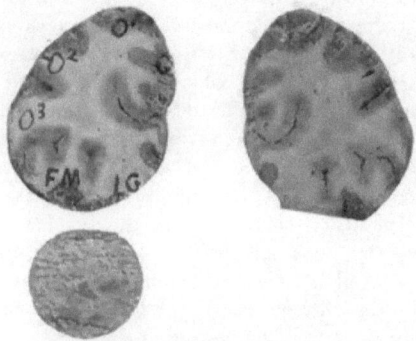

No. 36.

C—Cuneus.
$O^1\ O^2\ O^3$ — Occipital convs.
F M—Fusiform.
L G—Lingualis.

Figure 27 represents the upper surface of the cerebellum, and a section of the crura cerebri through the nates forward to the posterior perforated space. The nates are united by a commissure above the iter. The roof of the iter is arched, and its floor is formed by the fasciculi teretes, which are divided in the middle line by the central fissure. A section of the raphe dividing the cruri cerebri extends from the median fissure of the iter to the posterior perforated space. Surrounding the iter is a layer or tube of grey matter, which is the continuation of the grey columns of the spinal cord and the grey matter of the floor of the fourth ventricle. Downward and outward from the grey matter around the iter, in succession upon each side, are sections of the processus, fillet, locus niger, and crusta. External to these parts, upon a line with the nates on each side, are the brachia. The third nerve enters the inner side of the crus cerebri, divides into striæ, which spread as they pass through the fibres of the processus to reach the grey matter of the iter. That portion of the section above the crustæ, including the locus niger, is called the tegmentum. The broad notch in front of the cerebellum embracing the tegmentum, is called the incisura anterior, and the notch behind, the incisura posterior. Between these notches is an elevated portion of the cerebellum, the superior vermiform process, or middle lobe of the cerebellum; its middle portion is called the monticulus, a small lobe in the incisura anterior is called the lobus centralis, and a few transverse convolutions behind are named the commissura simplex. The superior vermiform process is continuous behind with the inferior vermiform process of the middle lobe (Fig. 29). The lateral portion of the upper surface of the cerebellum contains two lobes; the square lobe in front, and the semilunar lobe behind.

Figure 27.

Figure 28—Is an anterior view of the cerebellum, showing its hilus and the relations of the parts transmitted by it, and also a section of the fourth ventricle. Above the hilus is the elevated middle lobe of the cerebellum, concave in front to embrace the crura cerebri, called the incisura anterior. In the middle of the notch is a small lobe, the lobus centralis, which is the anterior extremity of the superior vermiform process, and rests upon the upper surface of the valve of Vieussens. Beneath the valve is the cavity of the fourth ventricle, containing the nodule, which is the anterior extremity of the inferior vermiform process, and is in relation with the under surface of the valve of Vieussens. The nodule is connected on either side with the flocculus, by a thin membrane called the commissure of the flocculi. Beneath the nodule are the other parts comprising the inferior vermiform process, viz: the uvula, pyramid, and commissura brevis. Upon each side of the uvula are the amygdalae or tonsils, external to which, on either side, are: the flocculus, digastric, slender, and the posterior inferior lobes, of the lateral mass of the cerebellum. The lateral lobes are divided by a deep notch extending from the roof of the fourth ventricle to the posterior part of the cerebellum, the incisura posterior.

The parts transmitted by the hilus are from within outward: the valve of Vieussens, processes, restiform bodies, and the pons, composed of a deep and superficial layer of fibres. These parts compose the crura, or peduncles of the cerebellum, beneath which are the flocculi on each side. The convolutions and fissures of the cerebellum radiate from the hilus, and the great transverse fissure of the cerebellum is seen extending out laterally from each crus; and, being continuous behind, divides the cerebellum into superior and inferior lobes.

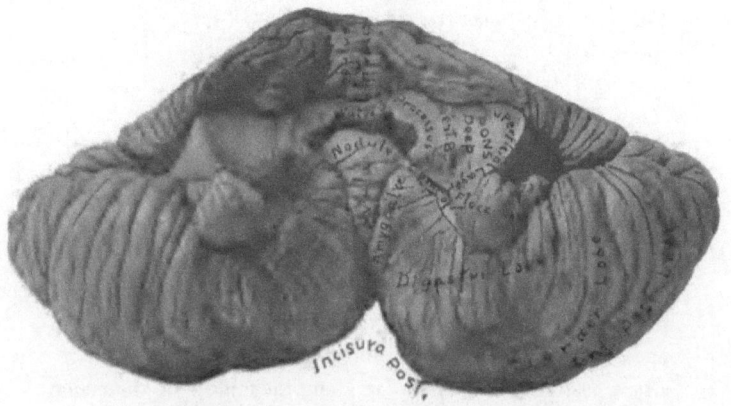

Figure 28.

Figure 29—Is a representation of the under surface of the cerebellum. Above and in front is the incisura anterior, upon each side of which is the flocculus. Behind is the incisura posterior, and extending along the middle of the under surface of the cerebellum, between the incisurae, is a deep furrow, at the bottom of which is the inferior vermiform process of the middle lobe of the cerebellum, the parts of which, from before backward, are: the nodule, uvula, pyramid, and commissura brevis. The inferior surface of the lateral hemisphere of the cerebellum contains from within outward: the tonsil, digastric, slender, and posterior inferior lobes. Behind the commissura brevis and the posterior inferior lobes, is the great transverse fissure of the cerebellum, which divides the upper from the lower surfaces. The separation of the sides of this fissure from behind uncovers the dentate bodies which adhere to the inferior lobes of the cerebellum. (See Fig. 15½.)

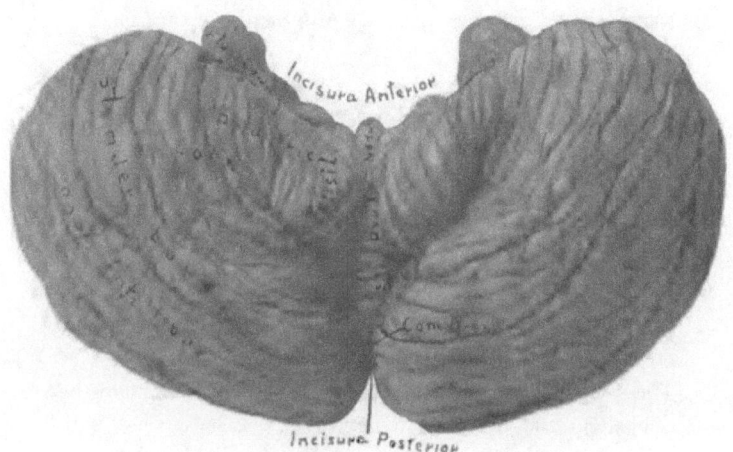

Figure 29.

Figure 30—Is a vertical section through the middle lobe of the cerebellum showing the arbor vitæ. The valve of Vieussens is seen to enter the hilus of the middle lobe of the cerebellum and divide into numerous branches; these subdivide into minute branchlets which are distibuted to the leaflets that line the sides of the fissures of the vermiform processes. The convolutions of the lateral lobes of the cerebellum radiate from the central lobe and are continuous with its convolutions and commissures.

Figure 31—Is a vertical section of the lateral lobe of the cerebellum, showing the convolutions and fissures of the hemispherical ganglion, the corona radiata, and dentate body. The processus enters the hilus of the cerebellum and terminates posteriously within the corpus dentatum.

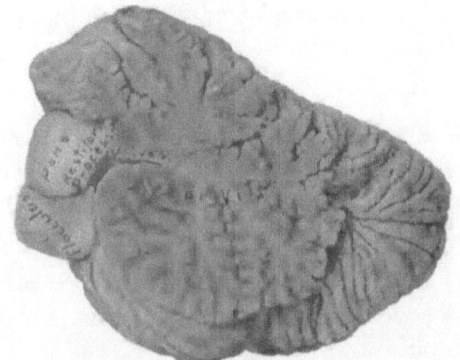

Figure 30.

Figure 31.

Figure 32—Is a transverse section of the crura cerebri midway between the upper border of the pons Varolii and the optic tract, and illustrates the crustæ and the parts which form the tegmentum.

Figure 33—Is a section of the crura cerebri, immediately above the pons Varolii, and shows the same parts that are seen in figure 32.

Figure 34—Is a section of the pons Varolii midway between its upper border and the emergence of the fifth nerves. It shows the fourth ventricle, fasciculi teretes, median fissure, processus, fillet, and the deep and superficial layers of the pons.

Figure 35—Is a transverse section of the pons Varolii on a level with the fifth nerves, the roots of which are seen on either side.

Figure 36—Is a section of the medulla oblongata through the olivary bodies and the restiform nuclei.

Figure 37—Is a section through the medulla oblongata at the decussation of its crossed pyramidal tracts. The anterior cornua of grey matter are interrupted, and the posterior cornua assume a rounded form.

Figure 38—Is a section of the spinal cord showing the arrangement of its grey and white colums, its commissure and central canal.

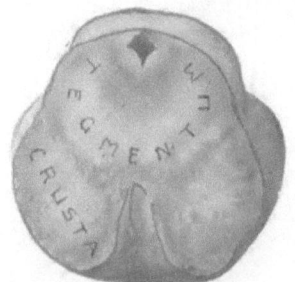

Figure 32.

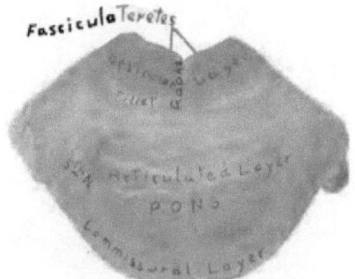

Figure 35.

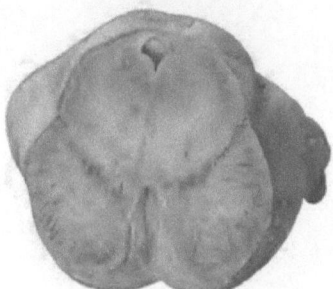

Figure 33.

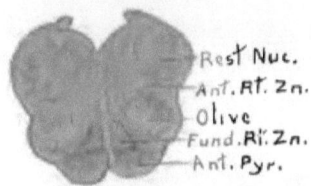

Figure 36.

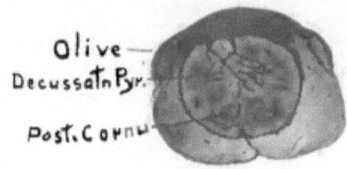

Figure 37.

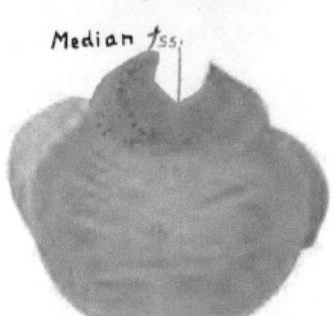

Figure 34.

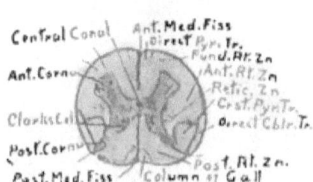

Figure 38.

INDEX OF CONTENTS.

	PAGE
Membranes of the brain	5
The cerebro spinal axis	8
The cerebrum	12
Dissection	18
Cerebellum	36
The pons Varolii	40
The medulla oblongata and spinal cord	51
Recapitulation of the tracts of the cerebro spinal axis	63
Central origin and relation to the cranial nerves	70
The cerebral hemisphere	106
Cerebral localization	128

INDEX OF ILLUSTRATIONS.

	FIGURES
Surface of the cerebrum	1 and 2
Corpus callosum and superior longitudinal commissure	3
Fornix and lateral ventricles	4
Velum interpositum and choroid plexus	5
Basal ganglia and superior surface of cerebellum	6
Base of the brain and lateral veiw of internal capsule	7
Posterior view of the intermedia	8
Lateral view of the intermedia	9
Posterior dissection of the intermedia	10
Dissections of crus cerebri, pons and medulla	11, 12, 13, 14
Fourth ventricle, processes and red nuclei	15
Relations of cerebellar peduncles and corpus dentatum	15½
External surface of the hemispherical ganglion	16
Internal surface of the hemispherical ganglion	17
Insula	18
External capsule and external longitudinal commissure	19
Lenticular nucleus and external longitudinal commissure and external capsule	20
Relations of crusta, fillet, thalamus and external longitudinal commissure	21
Relations of anterior commissure, crusta, fillet, thalamus	22
Intrahemispherical view of the internal surface of hemisphere	23
Relation of the convolutions and fissures to the skull	23½
Cerebral localization, after Horsley	24
Cerebral localization, after Ferrier	25
Median section of the brain	26
Upper surface of the cerebellum	27
Anterior view and hilus of cerebellum	28
Inferior surface of cerebellum	29
Arbor vitæ	30
Section of lateral lobe, cerebellum	31
Section of crus cerebri, tegmentum and crusta	32
Section of crus cerebri above the pons Varolii	33
Sections of pons Varolii	34 and 35
Section of medulla oblongata	36
Section of decussation of anterior pyramid	37
Section of spinal cord	38
Shirty-six topographical sections of the brain	pages 137 to 173

INDEX.

Anterior commissure of third ventricle, p. 26; Figs. 6-13-14; Sec. 14-15.
Anterior median fissure, p. 9; Figs. 7-9.
Anterior peduncles of corpus callosum, p. 21; Figs. 7-26.
Anterior pyramids of medulla oblongata, pp. 57-59; Figs. 9-11.
Anterior root zone of spinal cord, pp. 47-57; Figs. 7-10 to 13.
Aperculum, p. 111.
Acquaduct of Sylvius, or iter, p. 27; Figs. 26-27-32-33.
Arachnoid membrane, pp. 5-7.
Arbor vitæ, p. 37; Figs. 26-30.
Arciform fibres of medulla, p. 60; Figs. 7-9.
Basal ganglia, pp. 17-29; Figs. 6 to 10; Sec. 10 to 18.
Basilar groove, p. 9; Figs. 9-27.
Bichat, fissure of, pp. 10-13.
Brachia, p. 28; Figs. 6-8-9-13-14.
Burdach's column, p. 50.
Calamus scriptorius, p. 48; Figs. 8-10.
Caudate nucleus, pp. 30-17-23; Figs. 4-6-7; Sec. 11.
Central canal of spinal cord, p. 10; Figs. 8-10.
Central origin of cranial nerves, p. 70.
Cerebrum, pp. 8-12.
Cerebrum, superior surface of, p. 12. Figs. 1-24-26
Cerebrum, inferior surface of, p. 12; Figs. 2-7.
Cerebrum, structure of, p. 14.
Cerebral hemisphere, p. 106; Figs. 16-17.
Cerebral localization, p. 128; Figs. 23½-24-25.
Cerebellum, pp. 8-36; Figs. 6-7-27 to 31.
Cerebro-spinal axis, p. 8.
Choroid plexus, pp. 24-28-7. Figs 4-5
Choroid plexus of third ventricle, p. 25.

Claustrum, pp. 17-34; Sec. 10 to 18.
Clavate nuclei, p. 48; Figs. 8-10.
Commissure of gyrus fornicatus, p. 111; Fig. 3-23.
Comparison of cerebrum and cerebellum, p. 38.
Connection of superior and inferior peduncles of the cerebellum with the corpus dentatum, p. 38.
Convolutions of hemispherical ganglion, p. 111.
Cordon, pp 111-107; Fig. 17.
Cornua of lateral ventricles, p. 22; Fig. 5
Corona radiata cerebrum, pp. 16-18; Figs. 3 to 6.
Corona radiata, cerebellum, p. 25.
Corpora albicantia, p. 15; Figs. 7-9.
Corpora geniculata, pp 28-29; Figs. 7-8-9
Corpora quadrigemina, p. 27; Figs. 6-8-9.
Corpus callosum, p. 18 13-16; Figs. 3-8-23-26
Corpus fimbriatum, pp 23-24; Figs. 4-5.
Corpus dentatum, p. 37; Figs. 8 to 15½.
Crests of corpus callosum, pp. 18-23, Figs. 3-8.
Crossed pyramidal tracts, pp. 56.
Crusta, pp. 15-34; Figs. 2-7 to 12.
Decussation anterior pyramids of medulla, Figs. 7-11 to 14.
Decussation corpus callosum and internal capsule, p. 16; Figs. 3-8; Sec. 18-19.
Direct cerebellar tract, p. 56.
Direct pyramidal tract column of Türck, p. 55.
Dissection, p. 18.
Dura mater, p. 5.
Eminentia collateralis, p. 22; Fig 5.
External basal ganglion cerebrum, p. 33; Figs. 9-12-20.

NOTE.—Sec. signifies Topographical Sections.

INDEX—Continued.

External basal ganglion cerebellum, p. 39.
External capsule, pp. 17-33; Figs. 7-9-19; Sec. 10 to 18.
External longitudinal commissure, pp. 19-20.
Falx cerebri, p. 5.
Fascia dentata, p. 111.
Fasciculi teretes, pp. 44-47; Figs. 10-15-25.
Fifth ventricle, p. 24; Figs. 4 to 8.
Fillet or lemniscus, p. 43; Figs. 10 to 14, 26.
Fillet fibres from fourth ventricle, p. 47; Figs. 11-12-13.
Fissures, anterior median, p. 9.
Fissures, longitudinal, p. 10.
Fissures, Bichat or transverse, pp. 10-13-23; Figs. 4-5.
Fissures, posterior median, p. 10.
Fissures, olfactory, p. 15.
Fissures, Sylvius, pp. 13-108; Figs. 2-7-16-24.
Fissures, Rolando, p. 108; Figs. 24-25-23½.
Fissures, calloso-marginal, p. 110; Fig. 17.
Fissures, parieto-occipital, p. 110; Fig. 17.
Fissures, calcarine, p. 110; Fig. 17.
Flocculi, p. 36; Figs. 7-28.
Foramen of Monro, p. 23; Figs. 4-5.
Foramen of Majendie, p. 64.
Formatio reticularis, p. 20.
Fornix, p. 23; Fig. 4.
Fourth ventricle, pp. 48-11; Figs. 8-10-13-15-26.
Frontal lobes, p. 12; Fig. 2.
Fundamental root zone, pp. 15-28-46-47-55; Figs. 8-9-11-12-13.
Fusiformis lobe of thalamus, p. 32; Fig. 6.
Ganglion of pons Varolii, p. 45; Figs. 34-35.
Geniculate bodies, p. 14; Figs. 8-9-11-12.
Genu of corpus callosum, p. 24; Figs. 4-5-6-26.
Genu of internal capsule, pp. 24-30; Figs. 4-7-8.
Globus pallidus, p. 33.
Gowers' sensory tract, p. 57.

Great longitudinal fissure, p. 10; Figs. 1-2-17-24-26.
Groove separating corpura quadrigemina, p. 10; Fig. 8.
Gyrus fornicatus, p. 18; Figs. 17-26.
Hemispherical ganglion cerebrum, pp. 17-106; Figs. 16 to 23.
Hemispherical ganglion cerebellum, p. 38; Fig. 15½.
Hilus of cerebrum, p. 13; Figs. 2-17.
Hippocampus major, minor, p.22; Fig.5.
Incisura cerebelli, p. 10; Figs. 27-29.
Infundibulum, p. 26; Figs. 7-9.
Insula, p. 17; Fig. 18.
Internal capsule, pp. 16-29; Figs. 4 to 14, 18 to 22; Sec. 10 to 19.
Internal layer corona radiata, p. 20.
Internal basal ganglia, p. 30; Fig. 6; Sec. 9 to 17.
Internuncial fibres, p. 19; Figs. 3, 18 to 22.
Interpeduncular space, pp. 9-26; Figs 2-7-9.
Iter, p. 11; Figs. 10-15-26-27.
Lamina cinerea, pp. 21, 26; Figs. 9-11-12.
Lateral ventricle, pp. 22-11; Fig. 4.
Layers of corona radiata. p. 20; Figs. 20-21; Sec. 25.
Layers of pons Varolii, p 45; Figs 9-11-12.
Lemniscus, see fillet, p. 43; Figs. 10 to 14, 26.
Lenticular nucleus, pp. 33-17; Figs. 7-9-12-20; Sec. 17.
Ligamentum dentatum, p. 6.
Lobes of cerebellum, p. 36.
Loci cœrulii, p. 49.
Locus niger, p. 45; Figs. 10-11-12 32-33; Sec. 20-21.
Longitudinal commissures, p. 20; Figs. 3-4-9-18 to 21.
Lyra, p. 24; Figs. 5-6.
Median fissure, p. 9.
Medulla oblongata and spinal cord; pp. 57-51; Figs 2 to 15.
Middle commissure third ventricle, p. 26; Figs. 6-26.
Membranes of the brain, p. 5.
Mixed lateral tract of spinal cord, p. 57.

INDEX—Continued.

Nates (see corpora quad, page 27.)
Nerves of Lancisi, p. 21; Fig. 3.
Nucleus of pons Varolii, p. 45; Figs. 34-35.

Occipital lobes, p. 13.
Olfactory nerve bulb and fissure, p. 12; Figs. 2-7.
Olivary bodies, p. 60; Figs. 9 to 14.
Olivary fasciculus, pp. 47-60; Fig. 12.
Optic commissure, p. 26; Figs. 2-7.
Peduncles of cerebrum, p. 14; Figs. 2-7-8-9-10-26-27.
Pes hippocampi, accessorius, p. 22; Fig. 5.
Pia mater, p. 6.
Pillars of pineal body, **p. 27; Figs. 6-8.**
Pineal gland, p. 27; **Figs. 6-8.**
Pison, p. 31; Fig. 6.
Pituitary body, p. 26.
Pons Varolii, p. 40; **Figs. 7-9 to 14.** Structure of, p. 42. Dissection of, p 44; **Figs. 11 to 14.**
Posterior median column, Gall's, p. 55; Figs. 8-10.
Posterior median fissure, p 10; Figs. 8-10.
Posterior pillars of fornix, p. 24; Figs. 4-5.
Posterior pyramids of medulla, p. 57; Figs. 8-10-15.
Posterior root zone, Burdach's column, p. 50.
Processus, pp. 15-28; Figs. 8 to 15-27-28.
Pulvinar, p. 32; Fig. 6.
Puncta vasculosa, p. 19; **Fig. 3.**
Putamen, p. 33.
Raphe of cerebral peduncles, p. 14; Figs. 26-27 33. Oval opening in, p. 43; Fig. 26.
Raphe of corpus callosum, p. 21; Fig. 3.
Recapitulation of tracts cerebro-spinal axis, p. 63.
Red nucleus, p. 32; **Figs. 10-13-15;** Sec. 19-20.
Restiform body, p. 61; Figs. 8 to 15.
Restiform nucleus, p. 40; Figs. 8-10-36.
Restiform system, **p. 62.**
Restiform triangle, **pp. 47-61; Figs. 9 to 13.**
Section of corpus callosum, p. 21; Fig. 26.

Septum lucidum, **pp. 22-23-24; Figs. 4** to 10-26.
Spinal cord, pp. **8-61.**
Splenium, p. 21; Figs. 4-5-26.
Striæ, longitudinales, p. 21; Fig. 3.
Striæ, transversæ, p. 49; Figs. 8-10.
Structure of cerebrum, p. 14.
Subarachnoid spaces, p. 6.
Subthalmic ganglion (red **nucleus),** p. 32; Figs. 10-13-15.
Superior longi**tudinal commissure,** p. 21; Figs. 3-23.
System, cerebello spinal, **p. 68.**
System, fillet, p. 67.
System, ganglionic, **p. 64.**
System, motor, **p. 66.**
System, organic and sensory, p. 68.
System, respiratory reflex, p. 68.
System, ventricular, p. 63.
System, visual **reflex, p. 66.**

T**ænia** semi circularis, **pp. 30-23; Figs.** 6-7-8.
Tegmentum, pp. 15-16-34; Fig. 32.
Temporal lobes, p. 12.
Tentorium cerebelli, p. 5.
Testes (see corpora quad.) p. 27.
Thalamus opticus, pp. 17-23-31; Figs. 6-8-10.
Third **ventricle, pp. 26-11;** Figs. 6-8-10-15-26.
Tracts of spinal cord, p. 55; Figs 7 to 15-38.
Transverse fissure of Bichat; p. 10; Fig. 26.
Triangle to illustrate aphasia, **p. 19.**
Tuber cinereum, p. 26; Figs. **7-9.**
Tuberculum Thalami, p. 32; **Fig. 6.**
Valve of Vieussens, pp. 10-28; Figs. 8- 10-26.
Vena Galeni, p. 25; Fig. 5.
Velum interpositum, pp. 10-25; **Fig. 5.**
Ventricle of corpus callosum, p. 18.
Ventricular system, pp. 10-63.
Vermiform processes of cerebellum, p. 36; Figs. 26-27-31.
Vertical depression of crus cerebri, pp. 14-15; Figs. 8-9.
Vesicular columns **of Clarke, p. 52;** Fig. 38.
Vesicular columns **of spinal cord,** p. 52.

FULLER ANATOMICAL CO.,

MANUFACTURERS OF

ANATOMICAL AND PATHOLOGICAL MODELS,

GRAND RAPIDS, MICH.

Casts made of soft tissues, tumors, etc., of natural size and shape and representing minute details.
The specimens returned without injury.
Plates taken direct from the object for the illustration of books and papers a specialty

PRICES OF MODELS OF THE BRAIN.

Student's set, of six pieces in a box, represented by figures 1 to 7 and 16 and 17.................................$15.00
Six lateral dissections of hemisphere on a placque........... 15.00
Dissections of the intermedia, six pieces, Figs. 8 to 14........ 25.00
Topographical sections natural shape of brain, in a box...... 25.00

☞ *Dr. Fuller's models of the brain were awarded the medal and diploma of the World's Columbian Exposition, and were purchased by the University of Chicago.*

www.ingramcontent.com/pod-product-compliance
Lightning Source LLC
Chambersburg PA
CBHW032139160426
43197CB00008B/708